Advanced Materials for Emerging Water Pollutant Removal

Chemistry in the Environment Series

Titles in the series:

1: Graphene-based 3D Macrostructures for Clean Energy and Environmental Applications
2: Metallurgical Slags: Environmental Geochemistry and Resource Potential
3: Functional Hybrid Nanomaterials for Environmental Remediation
4: Emerging Nanotechnologies for Water Treatment
5: Biological Treatment of Industrial Wastewater
6: Advances in Functional Separation Membranes
7: Persulfate-based Oxidation Processes in Environmental Remediation
8: Advanced Ozonation Processes for Water and Wastewater Treatment: Active Catalysts and Combined Technologies
9: Environmental Nanopollutants: Sources, Occurrence, Analysis and Fate
10: Photo- and Electrochemical Water Treatment: For the Removal of Contaminants of Emerging Concern
11: Ionic Liquids for Environmental Issues
12: Novel Materials and Water Purification: Towards a Sustainable Future
13: Microbial Bioremediation and Multiomics Technologies for Sustainable Development: Recent Trends
14: Advanced Materials for Emerging Water Pollutant Removal

How to obtain future titles on publication:
A standing order plan is available for this series. A standing order will bring delivery of each new volume immediately on publication.

For further information please contact:
Book Sales Department, Royal Society of Chemistry, Thomas Graham House, Science Park, Milton Road, Cambridge, CB4 0WF, UK
Telephone: +44 (0)1223 420066, Fax: +44 (0)1223 420247
Email: booksales@rsc.org
Visit our website at books.rsc.org

Advanced Materials for Emerging Water Pollutant Removal

Edited by

Pei Sean Goh
Universiti Teknologi Malaysia, Malaysia
Email: peisean@petroleum.utm.my

Devagi Kanakaraju
Universiti Malaysia Sarawak, Malaysia
Email: kdevagi@unimas.my

Anwar Iqbal
Universiti Sains Malaysia, Malaysia
Email: anwariqbal@usm.my

and

Ahmad Fauzi Ismail
Universiti Teknologi Malaysia, Malaysia
Email: fauzi.ismail@gmail.com

ROYAL SOCIETY OF CHEMISTRY

Chemistry in the Environment Series No. 14

Print ISBN: 978-1-83767-117-5
PDF ISBN: 978-1-83767-542-5
EPUB ISBN: 978-1-83767-543-2
Print ISSN: 2516-2624
Electronic ISSN: 2516-2632

A catalogue record for this book is available from the British Library

The Royal Society of Chemistry is a charity, registered in England and Wales, Number 207890, and a company incorporated in England by Royal Charter (Registered No. RC000524), registered office: Burlington House, Piccadilly, London W1J 0BA, UK, Telephone: +44 (0) 20 7437 8656.

For further information see our website at www.rsc.org

Printed in the United Kingdom by CPI Group (UK) Ltd, Croydon, CR0 4YY, UK

Preface

Water scarcity is an increasing issue on every continent. Along with climate change, the global issue is escalated by urbanization, population growth, and pollution. According to the World Bank report, 80% of wastewater enters water bodies without being adequately treated. The release of vast amounts of industrial and domestic wastewater results in the pollution of the water environment and ecosystem. Proper treatment of these wastewaters allows their reutilization for many purposes and applications including industrial processes and irrigation. Advanced materials are known to be promising tools to alleviate the impacts of these issues. In the last decade, tremendous efforts have been made to progress in the field of synthesis and application of advanced materials especially for use in environmental remediation. Advanced materials including nanomaterials and biomaterials can be used to remove pollutants from water and air. A wide range of advanced materials can be prepared through affordable, energy-efficient approaches. They can also be easily retrofitted in existing wastewater or air filtration systems.

This edited book, *Advanced Materials for Emerging Water Pollutant Removal*, explores the potential of advanced materials to deal with the various kinds of pollutants found in water bodies. This book aims to bring together the ideas and innovations of researchers working in the field, and provides a detailed overview of the development of various functional advanced materials for the removal of emerging pollutants. This edited book comprises eight chapters that focus on the synthesis, characterization, and modification of nanomaterials as well as their applications and evaluation of their performance. The engineering of materials through innovative synthesis and modification approaches allows the fine-tuning and optimization of materials in terms of their functionality and efficiency. With the ever-growing threats of

Chemistry in the Environment Series No. 14
Advanced Materials for Emerging Water Pollutant Removal
Edited by Pei Sean Goh, Devagi Kanakaraju, Anwar Iqbal and Ahmad Fauzi Ismail

Published by the Royal Society of Chemistry, www.rsc.org

conventional and emerging pollutants in our water, the advancements made over the past decade could serve as a catalyst in revolutionizing efforts in environmental remediation.

We extend our heartfelt gratitude to all the contributors for their expertise and dedication in completing the chapters. Their insights are substantial in shaping the contents of this edited book. We would also like to thank the publisher for their support throughout the preparation and production processes. Lastly, by harnessing the advantages of a broad range of advanced materials, we aim to not only address the alarming pollution and water scarcity issues, but also contribute to the Sustainable Development Goals particularly Sustainable Development Goal 6: Clean Water and Sanitation. It is hoped that this edited book will provide inspiration and guidelines to propel the research community towards a cleaner and more sustainable future.

Pei Sean Goh
Devagi Kanakaraju
Anwar Iqbal
Ahmad Fauzi Ismail

Contents

Chemistry in the Environment Series No. 14
Advanced Materials for Emerging Water Pollutant Removal
Edited by Pei Sean Goh, Devagi Kanakaraju, Anwar Iqbal and Ahmad Fauzi Ismail

Published by the Royal Society of Chemistry, www.rsc.org

CHAPTER 1

Preparation and Modification of New Functional Materials for Organic Pollutant Elimination

C. RIZZO, S. AMATA, G. EMMOLA, S. BUSCEMI, A. PACE* AND A. PALUMBO PICCIONELLO

Università degli Studi di Palermo, Dipartimento STEBICEF, Sezione di Chimica, Viale delle Scienze Ed. 17, 90128 Palermo, Italy
*E-mail: andrea.pace@unipa.it

1.1 Introduction

Water is one of the most important resources in the world to ensure a safe life for billions of people. Considering the increasing population and the rapid industrialization and urbanization, the demand for water is constantly growing, and the need for drinking water, as well as fresh water for industrial facilities, is becoming a major issue.

According to the Food and Agriculture Organization (FAO), due to water pollution and misuse, billions of people will still lack access to basic water services in 2030 unless progress quadruples in speed.[1] In order to resolve this problem, novel and effective methods for removing pollutants from water are needed, in particular for emerging pollutants such as pharmaceuticals, perfluorinated compounds (PFC), solvents and nano-plastics.[2]

Processes for treatment of contaminated ground water include many techniques such as micro/ultra/nano filtration, centrifugation, (co)precipitation,

Chemistry in the Environment Series No. 14
Advanced Materials for Emerging Water Pollutant Removal
Edited by Pei Sean Goh, Devagi Kanakaraju, Anwar Iqbal and Ahmad Fauzi Ismail

Published by the Royal Society of Chemistry, www.rsc.org

(photo)oxidation, reverse osmosis, *etc.* which have been thoroughly discussed in the literature.[3–5]

Adsorption can efficiently be applied for the removal of different pollutants (organic, inorganic, and biological) and can be considered the cheaper way to produce purified water. In general, the adsorbate (pollutant) containing solution is kept in contact for a certain time, passed through, shaken, or stirred with the adsorbent, and the former adheres onto the surface of the adsorbent, *via* chemical or physicochemical interactions, under specific conditions, until equilibrium is established. Each type of adsorption depends on the way the adsorbent and adsorbate interact. Chemisorption is irreversible, because of the strong chemical bonds established between the adsorbate and the adsorbent surface. In contrast, physisorption is reversible, due to weak intermolecular physical forces involved (π–π and dipole–dipole interactions, Van der Waals forces, hydrogen bonds), between the adsorbate and adsorbent.[6] In developing a water remediation strategy, tests are usually performed at a lab-scale with single or multipollutant systems to simulate wastewater model systems. In this context, unselective adsorbents based on cheap or readily available materials can be useful to remove most pollutants reaching removal efficiencies close to complete removal. In some cases, the water resulting from the first round of remediation could be subjected to further cycles reaching the desired level of removal and water purity. However, when the residual pollutants are of a different chemical family with respect to the main ones, using the same type of adsorbent for subsequent cycles may not be a suitable solution for the complete removal of impurities. In these cases, the degradation of residual pollutant could be performed by several approaches that might include photocatalysis or other advanced oxidation processes (AOPs). Nevertheless, in case of persistent pollutants that are recalcitrant to most degradation processes or in the case of degradation that does not mineralize the pollutant but produces smaller/different by-products, the removal of such substances should be approached by specific adsorbing systems. This last step of water purification, that could be defined as fine water remediation (FWR), requires methods that are economically and energy-wise competitive with "simple" distillation and that are based on the availability of adsorbents for specific classes of residual persistent pollutants.

For this reason, many different approaches have been attempted, and different materials were synthesized depending on the nature of the pollutants to be removed. Organic materials such as biomass derived polymers, metal and covalent organic frameworks, carbon nanomaterials (graphene oxide, carbon nanotubes), and hybrid materials have been recently synthesized or functionalized for pollutant removal. In this chapter selected recent examples will be presented to give a general overview of this interesting and wide research field.

1.2 Biopolymers

The scientific community has shown increasing interest in biopolymers and their derivatives because of their unique characteristics, such as biodegradability and biocompatibility. Most biopolymer's monomer units are made up of repeating molecules of nucleic acid, nucleotides, amino acids, proteins, or saccharides.[7,8]

They can be classified into three categories: natural (polysaccharides like chitosan, alginate, cellulose, and proteins like gelatine, collagen), synthetic (made by chemical or biological modification of natural ones) and microbial biopolymers (produced through a fermentation process of monomers in bacteria and fungi like xanthan, cellulose).[9–11]

In terms of functional properties and manufacturing costs, biopolymers have created an ideal competition with their fuel fossil-based equivalents. In fact, research in this area is still ongoing with the goal of developing new biopolymers with ever-improving functional properties and ever-expanding uses. Biopolymers have recently found to have a variety of applications, from additives and bioplastics to personal hygiene, and medical products, whilst also offering the advantage of ecologic degradation.[12] Biopolymer applications for wastewater treatment are receiving more attention thanks to their properties and those of their composites. The expense of the clean-up procedure is decreased by using economical polymers as adsorbents. Because of their superior durability, high functionality, processing capabilities, and huge surface area, biopolymers are more effective at removing pollutants from the environment through adsorption.[13,14]

In this chapter, some biomaterials were selected for each class. The number of applications for cellulose, chitosan, alginate, and lignin has expanded in recent years, indeed, they are frequently found as by-products or waste and are becoming basic and freely accessible resources according to the principles of the circular economy.[15]

1.2.1 Cellulose and Its Derivatives

The distinctive macroscopic fibrous structure of cellulose is made up of a linear polymer chain formed by multiple D-glucose units connected by β-1,4-glycosidic linkages, and alternating sections of amorphous and crystalline zones. Several natural sources, including plants, seeds, cane, sand, and certain microbes, can yield highly pure cellulose.[16]

The raw materials balsa wood, loofah and pomelo peel have been combined with polyethyleneimine (PEI) to produce high-efficiency adsorbents for the removal of perfluorooctanoic acid (PFOA) from water. The three different starting materials were chosen for their easy modification and abundant presence as by-products, PEI was chosen because of the presence of a repeating amino group capable of binding the negatively charged residues of

PFOA.[17] Essentially, hydroxyl groups work as nucleophiles in almost all cellulose reactions, and this type of nucleophilic substitution reaction provides a greater range of functionalization possibilities. In this case epichlorohydrin, under alkaline conditions, produces an epoxy cellulose ether structure. On this intermediate, a further substitution occurs with the PEI (grafting approach), Figure 1.1.[18]

The maximum adsorption of PFOA is reached at pH 3 with a capacity of 279.3 mg g^{-1} for the material obtained from balsa wood. The systems were saturated for 4 h and the presence of salts promoted the adsorption through salting-out phenomena in addition to the electrostatic and hydrophobic interactions.

To remove polystyrene nanoparticles, cellulose's porous qualities and layered double hydroxides (LDHs) distinct positive charge were combined to create cellulose/MgAl LDHs composite beads. Van der Waals and electrostatic attractions dominated the interaction between nano-plastics and Mg/Al LDHs.[19] In this case the cellulose source comes from the *Pennisetum sinese* Roxb, an affordable and sustainable supply for polysaccharide. Using the entrapment approach, composite beads made of cellulose and LDHs have been produced.[20] The removal of nano-plastics using cellulose and hydrotalcites had a synergistic effect, with the maximum removal rate observed at a ratio of 1 : 1. The maximum activity is reached under acidic conditions, achieving 90% of efficiency. The efficient removal of nano-plastics is facilitated by the hydrotalcite's electrostatic attraction to the nano-plastics. Cellulose/LDHs composite beads were more successful in removing pollutants and they can be simply recovered and separated and are not toxic in the environment.[20]

The combination of polyaniline (PANI) and polypyrrole (PPy) with hydroxyehtylcellulose (HEC) is investigated for the removal of dyes (Methyl Orange and Rhodamine B) from aqueous media. PANI and PPy were

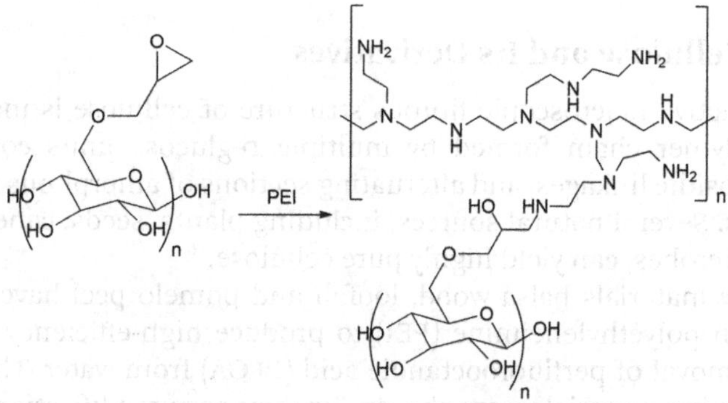

Figure 1.1 Cellulose functionalized with PEI synthesis.[17]

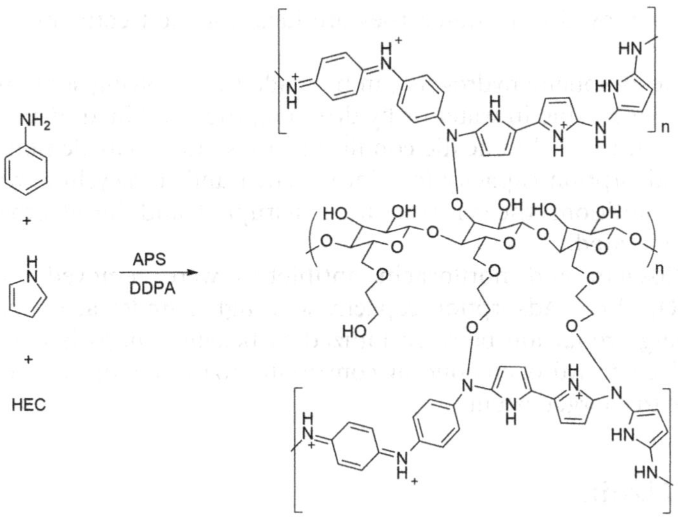

Figure 1.2 HEC-PANI-PPy polymer synthesis.[21]

chosen for the affordable cost, biocompatibility, non-toxicity, easy synthetic process (Figure 1.2), and significant surface area.[21]

1.2.2 Alginate

Alginate is a water-soluble linear polysaccharide derived from brown seaweed. It can collect metallic ions due to its surface which is rich in carboxylic and hydroxyl functional groups. Nevertheless, it is characterized by excessive rigidity and fragility, as well as weak elasticity and mechanical characteristics.

Chitosan combined with alginate solution forms an interpenetrated network, which, with the aid of a metal precursor (Fe^{3+}, Zr^{4+}, Al^{3+}), was concurrently transformed into an interconnected polymer chain mesh[22] by using two hybrid polysaccharides in a complex matrix distribution with Zr–Al sites forming a strong adsorbent for Cr^{6+} and F^- ions. The system works at pH = 4 and has a capacity of 315.86 mg g^{-1} for Cr^{6+}, and 52.75 mg g^{-1} for F^-. The mechanism of action is attributed to the ion exchange capacity and electrostatic interactions.

In recent years, novel materials have been developed by combining alginate with other synthetic or natural polymers, such as polyacrylic acid and chitosan, to improve mechanical and thermal stability and adsorption capabilities. Furthermore, numerous organic and inorganic–alginate materials have been produced and tested for the removal of antibiotics from wastewater.

Encapsulating or incorporating graphene oxide (GO) into an alginate-based hybrid material not only improves the mechanical and physical properties of the alginate, but also reduces GO nano-toxicity. Because hydrogels may be easily removed from water, they are ideal for the treatment of aqueous matrices.

Alginate–graphene hydrogels can be made by dissolving sodium alginate and graphene oxide in water or by dropping the combined dispersion into a $CaCl_2$ solution.[23,24] In acidic conditions, these materials demonstrate the highest adsorption capacity for ciprofloxacin and tetracycline, whereas in alkaline conditions, the gel structure is disrupted, and the adsorbed chemicals are liberated.

Ciprofloxacin and norfloxacin antibiotics were removed using alginate-Fe_3O_4 fibres. Adsorption capacity was higher under acidic conditions since charge repulsion between ionized carboxylic groups is reduced, and intermolecular hydrogen bonding contributes to the stability of the hydrogel structure to a larger extent.

1.2.3 Lignin

Lignin is a heterogeneous aromatic polymer found in nature, made up of cross-linked phenylpropanoid units (*e.g.* *p*-hydroxyphenyl, guaiacyl and syringyl units).[25]

For the removal of phosphate from water the application of sawdust functionalized with iron-binding ligands is interesting. The amount of phosphate that sawdust may bind is insufficient (0.3 g kg^{-1}). But because iron and phosphate have a great affinity for each other, the formation of iron–phosphate linkages presents a possible path for the creation of recyclable resins. The three-step, fully aqueous synthesis of ethylene diamine/Fe(III)-functionalized lignin produces a material that has shown remarkable efficacy (40 g kg^{-1}) and a recyclable nature as a phosphate-binding ligand.[26] Previously synthetized iron–chitosan composite has a capacity of 8.2 g kg^{-1}, but recently the price of this biopolymer has increased; therefore, lignin could be a cheaper option demonstrating better efficiency as well.

1.3 Metal and Covalent Organic Frameworks

When most pollutants have been removed from wastewater, the porosity of the adsorbent is one of the main properties that needs to be considered for residual pollutant removal. In this framework, metal–organic frameworks (MOFs),[27–29] a type of crystalline organic–inorganic hybrid compound and the corresponding covalent organic frameworks (COFs),[30] thanks to their high porosity, have strongly been applied for environmental remediation.[31,32] MOFs and COFs have demonstrated great efficiency in the removal of both inorganic and organic pollutants.[33,34] The reason behind the recent large use of these materials for environmental remediation lies in the easy fine tuning of their channels through pre- or post-synthetic approaches.[35,36] The channels can be varied in terms of size, shape, and functionality giving rise to a controlled host–guest chemistry. For all these reasons they have also recently been applied for the removal of some emerging pollutants, however, the impact of the synthetic approach of these materials is also important considering the environmental application.[37]

1.3.1 Removal of Biological Contaminants by MOFs

A lanthanum/palladium-metal organic framework (La/Pd-MOF) has been used to remove triclosan, an antimicrobial agent often used in detergents, soaps, toothpaste and cosmetics, becoming an increasingly dangerous presence for water purity.[38] La/Pd-MOF was easily synthesized mixing solutions of palladium and lanthanum nitrate to a solution of 2-methylimidazole.[39] The resulting MOF was characterized by an amorphous nature, that supported a high level of antimicrobial elimination effectiveness. After adsorption, the surface area and the pore volume of La/Pd-MOF decreased dramatically indicating the effective adsorption. The adsorption process was influenced by pH and a chemisorption process was identified which was involved in the overall process. In particular, the mechanism of interaction between the La/Pd-MOF and pollutant could involve π–π interaction, pore filling, H-bonding, or electrostatic interaction. In addition, the synthesized MOF adsorbent can be recyclable up to five cycles of adsorption–desorption.[40]

On the other hand, a Zn-MOF synthesized from an amide containing carboxylic acid *N*,*N*″-(3,5-dicarboxylphenyl)benzene-1,4-dicarbox-amide and 4,4′-bipyridine (Figure 1.3) has been recently used for sensing activities toward biomarker (2-methoxyethoxy)acetic acid and antibiotic tetracycline hydrochloride.[41] A MOF structure is formed by two-dimensional layers stacked to form a stable supramolecular structure with one-dimensional channels, the single unit is held together by the bond between carboxylate ligand and 4,4′-bipyridine with Zn(II) centers. Luminescence measurements indicated that the complex exhibited excellent multiple sensing activities.[41]

Another emerging pollutant is the endotoxin secreted from bacteria; cyanobacteria can cause severe health problems such as impairment of the immune system and toxic shock. To avoid its presence in water, a novel modified amino Cu-MOF was synthesized.[42] Firstly, the Cu-MOF was synthesized reacting $Cu(NO_3)_2$ and benzenetricarboxylic acid, to this preformed

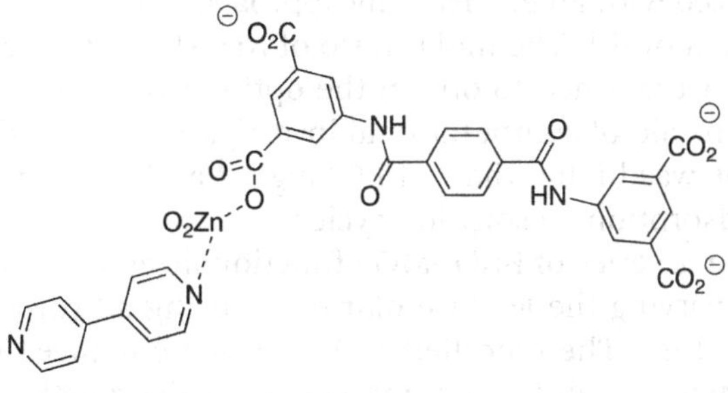

Figure 1.3 Single unit of Zn-MOF used for sensing.[41]

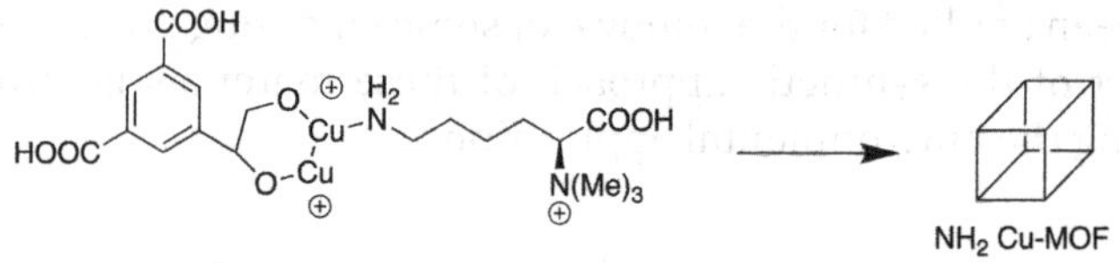

Figure 1.4 Structure of a single unit and schematic representation of cubic NH_2 Cu-MOF.[42]

crystalline MOF a post-functionalization reaction involving lysine to obtain an ammine tail was carried out (Figure 1.4).

X-ray and SEM analysis showed the cubic crystalline nature of modified MOF and this was used for the adsorption of endotoxin. Adsorption experiments performed in several conditions demonstrated great efficiency of MOF with over 90% of endotoxin degradation and indicated that the adsorption was driven by electrostatic, hydrophobic bonding and hydrophobic interactions held together by endotoxin and NH_2-Cu(MOF). Recycling tests also showed good reusability and stability of NH_2-Cu(MOF), with a decrease in removal efficiency of only 93.4–89.1% after five subsequent recycles.

1.3.2 Removal of Organic Pollutants by COFs and MOFs

To test the adsorption ability of materials against organic contaminants dyes are often used, for example an anionic sulfonate-functionalized covalent organic framework (COF) was applied. The selected COF was yielded by the Schiff-base condensation between 2,4,6-triformylphloroglucinol and 4,4′-diaminostilbene-2,2′-disulfonic acid under solvothermal conditions in the presence of sodium acetate (Figure 1.5). Thanks to the dense sulfonate groups in the periodic one-dimensional channels, in only 10 min the negatively charged COF showed a 99% uptake of multiple cationic dyes such as methylene blue, crystal violet, malachite green and janus green.[43] The efficiency was higher with respect to those of MOFs, also demonstrating good recyclability. In addition, COF presented low adsorption capacity for anionic dyes rendering it selective for the separation of cationic and anionic dyes.

Recently, a copper–gallic acid metal–organic framework (CuGA MOF) has been synthesized with an eco-friendly approach in order to use the material for basic dye removal.[44] The molar ratio of NaOH and GA, temperature, and reaction time, were tuned to obtain the optimal yield and crystallinity percentage with the aid of computational investigations. The adsorption capacity of this MOF was higher than 115.08 mg g^{-1} and this efficiency remained even after 3 adsorption–desorption cycles.

Furthermore, a series of Bi_2O_3-MOF functionalized with iron succinic acid was prepared varying the Fe : Bi molar ratios using a hydrothermal method at low temperature. The modified MOF showed enhanced properties than the pristine MOF overall for the photocatalytic degradation of the cationic organic dye, Rhodamine B. Modified MOF with molar ratios of Fe : Bi (1 : 0.1)

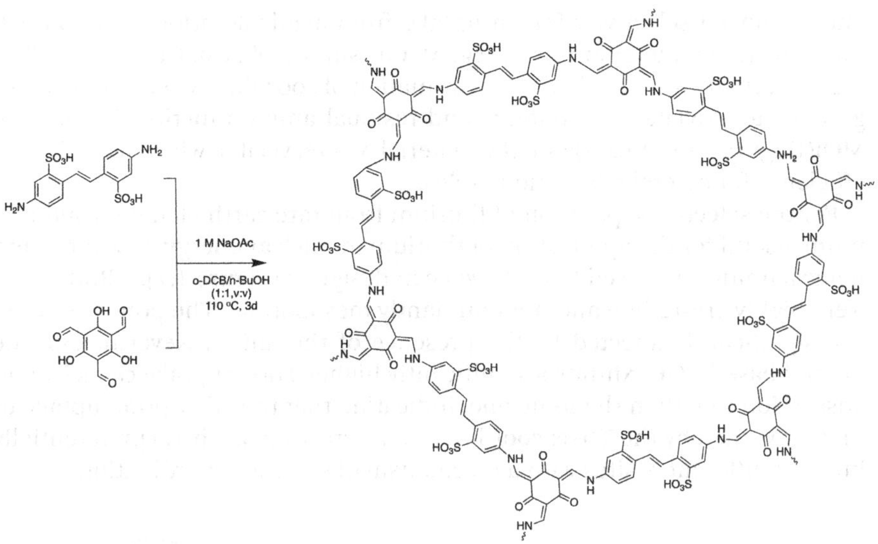

Figure 1.5 Synthetic approach to achieve sulfonate-COF. Adapted from ref. 43 with permission from Elsevier, Copyright 2023.

exhibited the highest degradation efficiency of 80% in 120 min under visible irradiation.[45] The degradation mechanism proposed that the introduction of Bi_2O_3 into the Fe-succinic acid framework, facilitated the charge carrier's separation at the interface, responsible for the enhanced photodegradation efficiency of the nanocomposite photocatalysts. In addition, trapping experiments of the reactive species revealed the order hydroxyl radicals > superoxide anion radicals > holes for their role in the degradation process. Indeed, a new eco-friendly synthesis of the composite iron–succinic frameworks was designed without the use of any toxic organic solvents, taking care of the overall processes from the synthesis of the material to the application for the cost effective and potential technology for purifying contaminated water.

1.3.3 Removal of Metal Contaminants by COFs and MOFs

Other emerging pollutants in addition to the common heavy metals are rare earth elements that have started to be extensively used in renewable energy technologies and electronic devices. In addition, the depletion of natural rare earth elements from mineral deposits has made selective recovery of rare earth elements from alternative sources essential. In this context, a chromium-based MOF was first synthesized and then modified with *N*-(phosphonomethyl)iminodiacetic acid for the selective recovery of Eu. The modified MOF was achieved following a hydrothermal process, where the pristine Cr-MOF bearing an ammino free functionality was modified with *N*-(phosphonomethyl)-iminodiacetic acid in the presence of *N*,*N*-dicyclohexylcarbodiimide under reflux.[46] The adsorbent showed a good adsorption capacity and enhanced selectivity for Eu uptake from acidic solution over competing transitional metal ions (Na, Mg, Al, Ca, Mn, Fe, Ni, Cu, Co, and Zn). The high selectivity was ascribed to the formation of coordinative complexes with grafted carboxylate, phosphonic, and residual amine functional groups of MOF (Figure 1.6). Once again, the material was recyclable without any loss in efficiency for several adsorption cycles.

For the selective separation of thorium from rare earth elements and uranium, useful for the application of thorium in nuclear energy, new ionic and neutral imidazole-based COFs have been designed (Figure 1.7).[47] Both COFs were highly crystalline and predominantly mesoporous. The pore sizes were not significantly affected by the presence of the anion. Nevertheless, the neutral based COF exhibits a significantly higher Th(IV) uptake capacity and adsorption rate than the ionic one, indicating that the adsorption uptake of Th is governed by the Th–N coordination interaction as Th(IV) preferentially binds at different N sites with a mechanism of selective coordination.

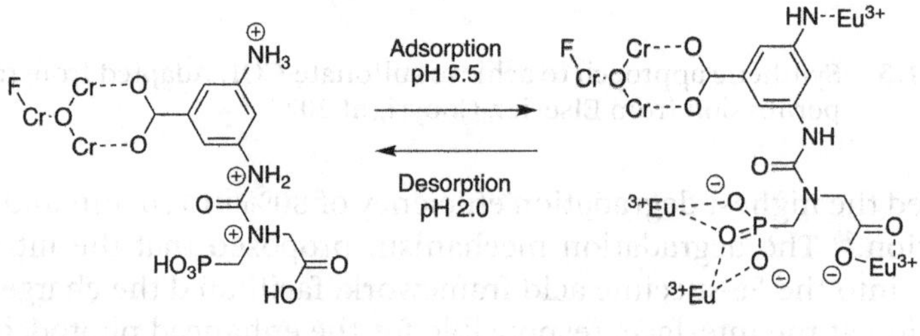

Figure 1.6 Schematic representation of adsorption–desorption mechanism on MOF.[46]

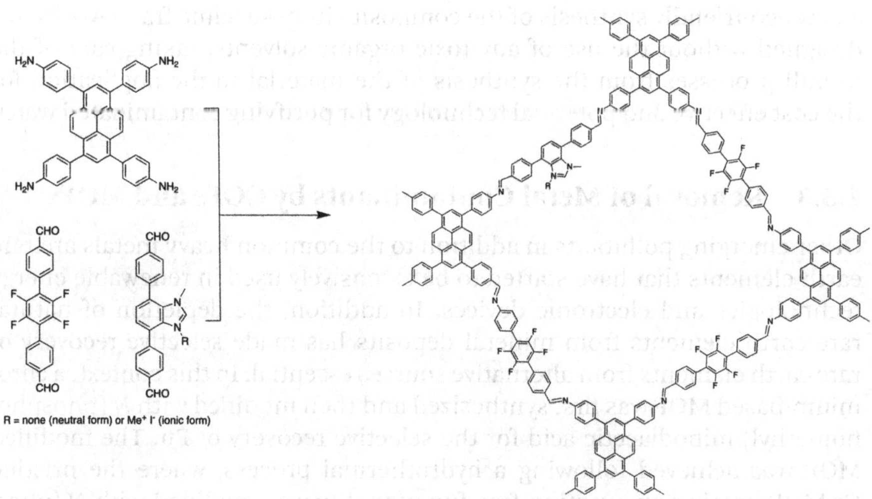

Figure 1.7 Synthetic approach to achieve ionic and neutral-COF.[47]

Several MOFs and COFs have been efficiently tested for the uptake of heavy metals from wastewater,[48,49] but recently a new water-stable MOF prepared from the combination of two different oxamide-based metallo-ligands derived from L-serine and L-methionine has been synthesized and applied for the efficient removal of both inorganic and organic pollutants.[50]

MOF is characterized by hexagonal channels bearing two types of flexible and functional tails ($-CH_2OH$ and $-CH_2CH_2SCH_3$) allowing the simultaneous and efficient removal of both kinds of pollutant: heavy metals, such as Hg^{2+}, Pb^{2+}, and Tl^{+} and organic dyes such as Pyronin Y, Auramine O, Brilliant green, and Methylene blue. This MOF can be used for the removal of many kinds of pollutants from water resources and can be recyclable, thus opening new perspectives for this emerging type of multivariant MOF.

1.4 Functionalized Carbon-based Nanomaterials

Carbon-based nanomaterials (CMs) such as graphene-based materials, carbon nanotubes, nanofibers, and fullerene, are compounds whose structure, mainly represented by carbon atoms, allows them to self-assemble into many different two- and three-dimensional structures, each with its own properties. CMs have recently attracted attention as pollutant-removing agents thanks to their excellent physicochemical and mechanical stability, the presence of multiple active surface sites, the high biocompatibility and large possibility of functionalization,[6,51] and lastly, whether the pollutant in charged, ion-exchange can be performed by giving the CM the opposite charge.[52]

In batch processes CMs have shown high capability in binding small pharmaceutical molecules, allowing them to be used in the purification of wastewater. Granularly activated carbon F-400, carbon nanotubes, and nanofibers have shown high adsorption rates of 242 mg g^{-1} for carbamazepine and 264 mg g^{-1} for ciprofloxacin.[53]

Graphene oxide (GO) has been able to remove 500 mg g^{-1} of diclofenac and 3709 mg g^{-1} of sulfamethoxazole. The mechanism for the adsorption was due to hydrogen bonds, hydrophobic and π–π interactions. Furthermore, an adsorption value of 149.4 mg g^{-1} of 17β-estradiol has been calculated for GO. This happens due to the hydrogen bonds established between hydroxyl groups and π–π interactions because of the presence of the aromatic ring in these hormone (Figure 1.8).[54]

In addition, biochar, a porous carbonaceous material produced by pyrolysis of biomass, has demonstrated an adsorption efficiency for ibuprofen of 167 mg g^{-1} at pH = 2. However, if the biochar is functionalized with iron chloride, becoming magnetic, it can be easily recovered after absorption but its efficiency decreases to 140 mg g^{-1} at pH = 4.[55]

Another example occurs in polymeric magnetic functionalized GO composites, which has been used for the removal of chlorinated pesticide such as chlorpyrifos and, moreover, exceptional adsorption capacities for bisphenol-A, with a value of 182 mg g^{-1} at pH range of 2–4.[56] This can be due to the formation of H-bonds between –OH/–COOH groups (Figure 1.9).

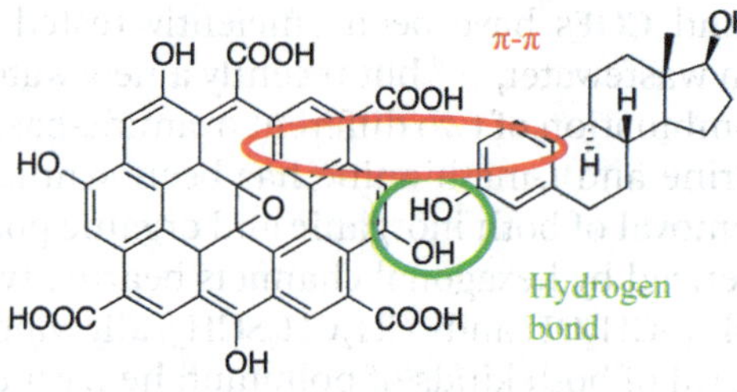

Figure 1.8 Interactions between GO and 17β-estradiol.[54]

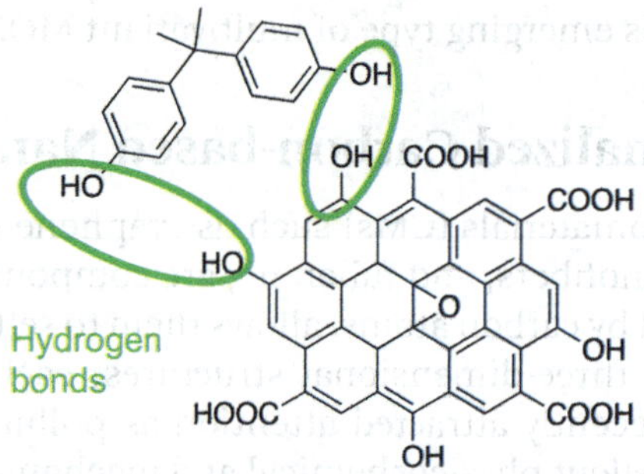

Figure 1.9 Interactions between graphene and bisphenol-A.[56]

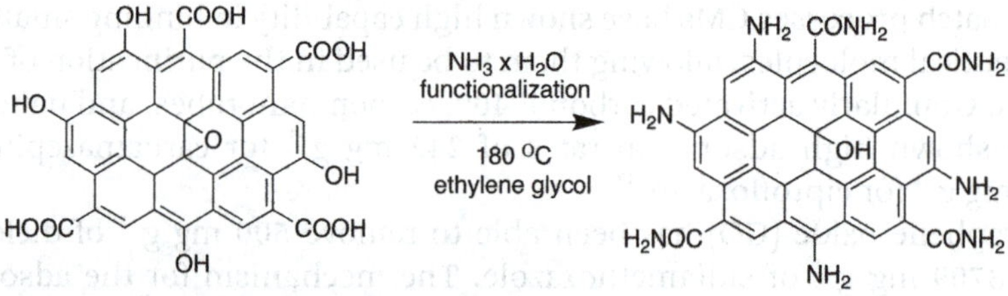

Figure 1.10 Structural representation of NH_3-GO.[57]

Verma *et al.* performed an ammonia functionalization of the GO sheets, by redispersing 100 mg of GO in 40 mL of ethylene glycol, adding 0.1 μL mg^{-1} of liquid ammonia until the dispersion color turned black. The resulting NH_3-GO (Figure 1.10) was able to remove dyes from an aqueous solution in relevant percentages: 56.2% of Basic Blue 41, 52.3% of Methyl Orange and 28.3% and of 4-nitrophenol.[57,58]

Sajid *et al.* investigated the possibility of using CMs to remove microplastics. Because they are less than 3 μm microplastics are difficult to remove from aqueous media, it is therefore crucial to design non-toxic and biodegradable adsorbents with the ability to remove small size MPs, to avoid the phenomenon of secondary pollution.[59] Graphene oxide (GO) sponges have been developed as materials for the efficient removal of microplastics

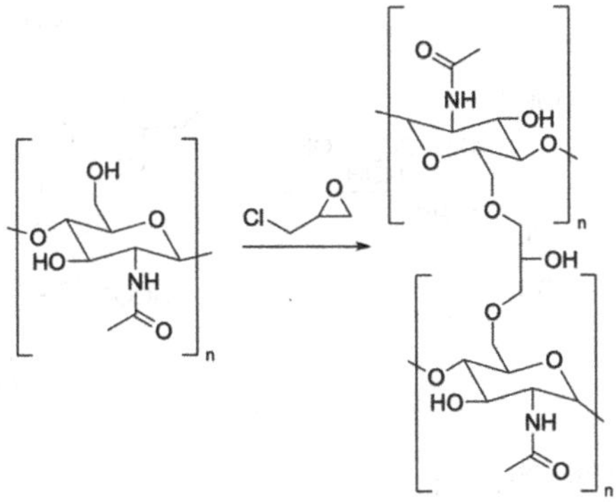

Figure 1.11 Chitin cross-linking mechanism, through epichlorohydrin.[60]

from aqueous solutions. These sponges are made by incorporating GO powder into a chitin hydrogel (a non-toxic and biodegradable polysaccharide), which increases their strength and enhances their pollutant removal performance. The procedure involves crafting a chitin hydrogel through a specific blend, establishing a dual-crosslink structure (Figure 1.11)[60] with solid GO homogeneously distributed, eliminating residues, and transforming the hydrogel into sponges *via* cryogenic treatment and vacuum freeze-drying. The overall method demonstrates a methodical and controlled approach to produce sponges with the desired characteristics from hydrogels. The presence of functional groups on GO contributes to the removal of positively, negatively, and neutrally charged pollutants (such as polystyrene (PS), PS-COOH, PS-NH_2). The chitin-based sponges created with GO have shown a removal efficiency of 71.6–92.1% for different types of MPs. The adsorption mechanism involves hydrogen bonding, π–π interactions, and electrostatic interactions.[54,59]

With the aim of increasing the sp^2 clusters in GO to enhance π–π or other hydrophobic interactions for increasing adsorption of organic compounds, another material has been synthetized by the introduction of a lipophilic functional group on GO. In detail, GO was functionalized by the formation of the covalent bonds between carboxyl groups of GO and amine groups of 9-aminoanthracene to obtain GO-9-AA (Figure 1.12). This material was utilized for the removal of naphthalene, acenaphthylene, and phenanthrene.[61]

1.5 Hybrid Materials

Hybrid materials are commonly employed due to their ability to combine the advantages of organic frameworks (easy functionalization and modification, bio-compatibility, and renewability) and inorganic scaffolds (easy synthesis, stability, and durability).[23,62] The hybrid materials can be classified depending on their obtainment method as: (i) *Structurally Hybridized Materials* (SHM) if the organic–inorganic parts are composites that interact through

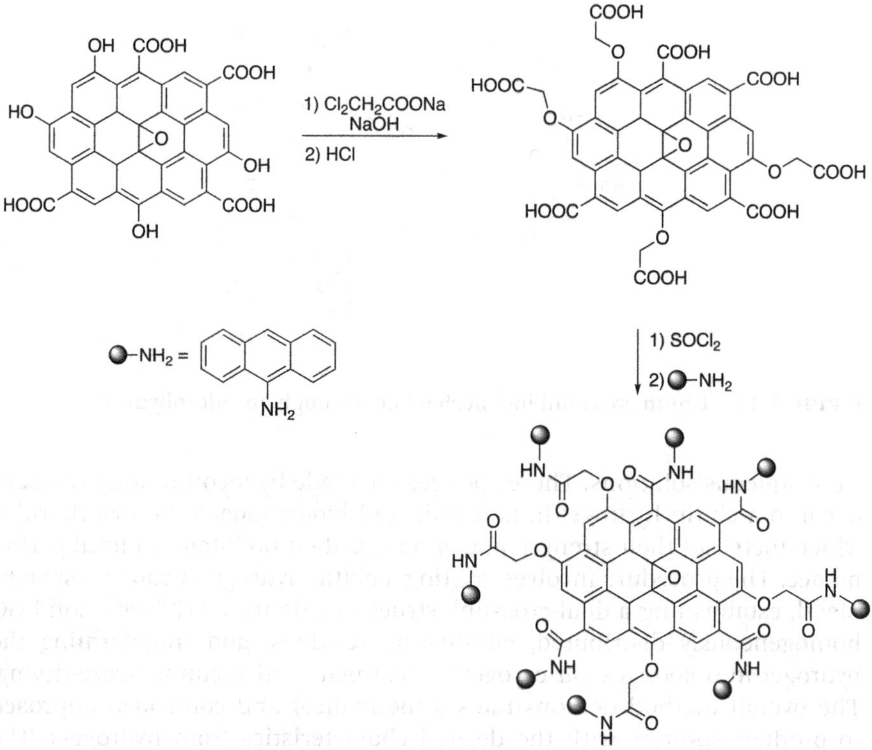

Figure 1.12 GO functionalization through 9-aminoanthracene.[61]

non-covalent bonds; (ii) *Materials Hybridized in Chemical Bonds (MHCB)* if the presence of covalent bonds between the organic–inorganic parts can provide new properties to the material. In recent years many different materials have been synthesized and currently the most common building blocks are bio-compatible organic polymers such as chitosan, gelatine, alginate, cellulose, carboxymethylcellulose (CMC), *etc.*

Concerning the inorganic portions, minerals, clay, and oxides, such as silica, have been preferred for their low cost and toxicity. The use of metal organic frameworks (MOF) or carbon-based materials such as graphene oxide (GO) is readily expanding in recent years. Below we will present some recent examples focusing on the employed inorganic building blocks and the different synthetic strategies.

1.5.1 Silica and Metal Oxide Based Hybrids

Silica and/or metal oxides are widely used as pollutant adsorbents due to their high surface area, defined and tuneable pore size, thermal stability, and availability.[63,64] Streide Machado and co-worker[65] designed a silica–chitosan composite crosslinked with glutaraldehyde with promising properties for

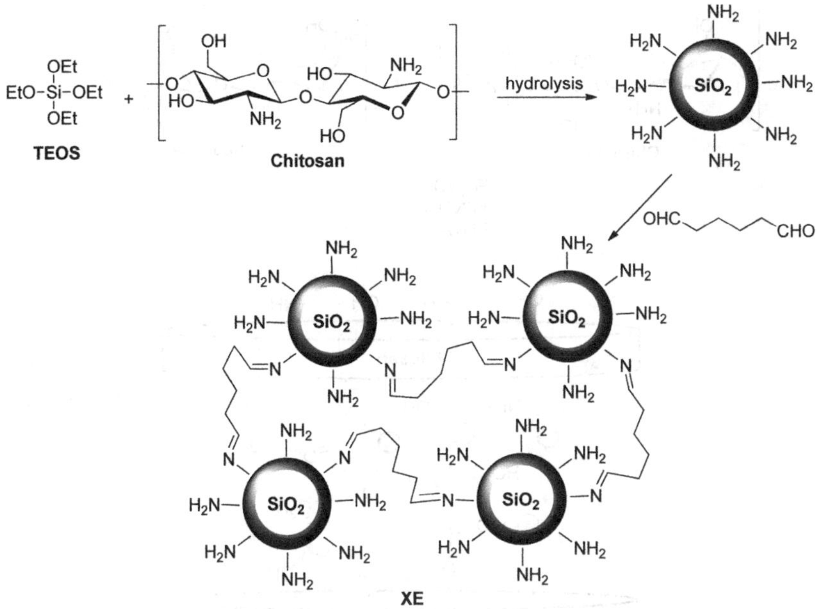

Figure 1.13 Synthesis of silica particles coated with chitosan and glutaraldehyde as crosslinker.[65]

the removal of diclofenac from water. The synthetic process starts with the hydrolysis of tetraethoxysilane (TEOS) in the presence of 1% chitosan. The obtained silica particles, coated with the biopolymer, were crosslinked with glutaraldehyde, with different ratios, thus forming xerogels named XE presenting covalent imino-bond (Figure 1.13).

The hybrid was characterized by means of FT-IR, BET, SEM, and elemental analysis. The adsorption isotherms revealed a high adsorption capacity of 237.8 mg g^{-1} for diclofenac. Notably, due the presence of free carbonyl groups material XE can also covalently link the amino moiety of the adsorbed drug. This feature was not previously observed with other similar composites and it is quite interesting for the prospective removal of other emerging contaminants.

Bionanocomposite-based aerogel (BNC-AG) is an α-FeOOH and γ-AlOOH-based nanocomposite formed within a chitosan and agarose aerogel reinforced with ethylenediaminetetraacetate dianhydride (EDTAD) and can be used for membrane filtration of multi-polluted water.[66] Fe–Al oxyhydroxide nanocomposite (NC) was formed *in situ* on preformed chitosan/agarose aerogel, obtained by means of the freeze-drying method, by dipping in the presence of EDTAD, which is hydrolysed and acts as a reinforcement of the membrane through covalent bonding with the hydroxyl groups of the polymers (Figure 1.14).

The microporous structure of the membranes was evidenced with FE-SEM and the material was investigated with multiple techniques: PXRD, FT-IR,

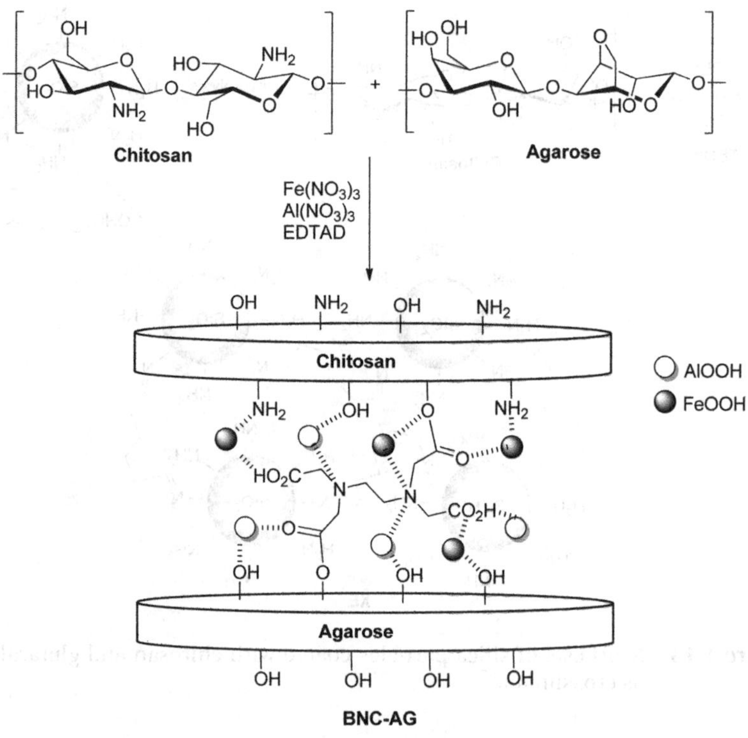

Figure 1.14 Synthesis of bio-nanocomposite-based aerogel (BNC-AG) membrane.[66]

BET, and XPS. The membrane can act as a filter for the removal of dyes, emerging pollutants, arsenate, and fluoride. In particular, the removal of ciprofloxacin, peptide hormone oxytocin, and non-ionic surfactant Triton X 100 from water was 91%, 89%, and 96%, respectively. Also, inorganic pollutants can be effectively removed, in fact, 93% As(v) and 99% F can be retained. The membrane can work under continuous flow and can be recycled up to 50 times with low efficiency loss after acidic washing, therefore it can be considered to be of practical application for wastewater purification.

1.5.2 Metal–Organic Framework Hybrids

MOF are MHCB materials with the great advantage of a regular and highly porous structure that ensures good adsorption properties.[67,68]

The advantage of using MOF is clearly shown in a recent example, surface functionalized metal–organic framework and biological macromolecule (SFMOF/BM) bio-composite was synthesized starting from MIL-53(Fe), a common MOF synthesized using the hydrothermal method. Surface modification with aminopropyl trimethoxysilane (APTMS) of the formed MOF was followed by mixing with chitosan (Figure 1.15).[69]

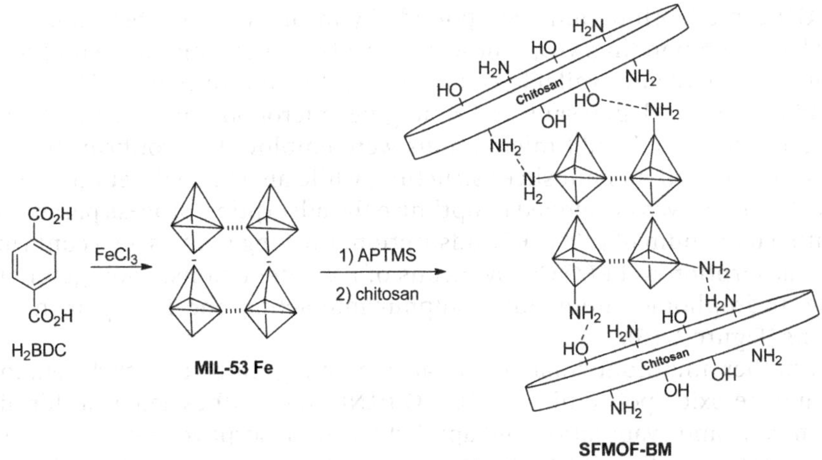

Figure 1.15 Synthesis of surface functionalized metal–organic framework and biological macromolecule (SFMOF/BM) bio-composite.[69]

The obtained SFMOF/BM composite was studied for adsorption capacities toward dyes such as Direct Red 23 (DR23) and emerging pollutants such as tetracycline and doxycycline. The adsorption capacity of SFMOF/BM was 388 and 264 mg g^{-1}, for tetracycline and doxycycline respectively. The saturated adsorbent can be recycled five times and can be considered a promising adsorbent for water treatment applications to remove emerging pollutants.

Other significant recent examples are sodium alginate (SA), MXene, MIL-101(Fe), composite hydrogel beads,[70] efficiently employed for the removal of naproxene in fixed-bed adsorption columns. Analysis of relevant parameters predict the potential design of an industrial-scale column. The regeneration of composites involved the indirect electrooxidation of the materials in the presence of OH radicals or indirect oxidation in the presence of active chlorine species.

A novel adsorbent formed combining poly(acrylamide-*co*-acrylic acid)/chitosan (P(AM-*co*-AA)/CS) composite with Zr-based MOF-808, was developed for the adsorption of U(VI) from seawater, through a rational design of the hybrids aiming to obtain a low swelling ratio in water reducing the solvent penetration filling the copolymer with CS.[71]

In fact, the adsorbent exhibited good mechanical properties because of the hydrogen bonding and covalent cross-linking between CS, AM, and AA and the coordinate bond between MOF-808 and P(AM-*co*-AA). The adsorption capacity of the adsorbent is 159.56 mg g^{-1} at pH = 8.0 also showing a U(VI) removal up to 95% in the presence of other ions.

1.5.3 Carbon-based Hybrids

Graphene, GO and other carbon-based compounds are effective nanomaterials for the removal of organic pollutants, heavy metals, anions, and bacteria, due to their high surface area and adsorption capacity.[58,72–74]

GO can adsorb organic compounds by means of π–π interactions and/or hydrogen bonding. GO, synthesized *via* Hummers' method, composites with commercially available cellulose with the aim to remove Flupentixol (FPLL), an antidepressant and emerging micropollutant.[75] XRD, Raman spectra, SEM and TEM micrographs were employed to confirm the presence of GO and its nanosheet structure, while an artificial neural network (ANN) model was developed to optimize the adsorption process parameters with a maximum of (99%) FPL adsorption. Docking studies also confirmed the absorption of FPL@GO by means of π–π interactions, hydrogen bonding, hydrophobic interactions, sulphur interaction and lone pair interactions (Figure 1.16).

Another multifunctional material containing carboxymethyl cellulose/graphene oxide/polyaniline (CMC/GO/PANI) was synthesized as a thin film by mixing and evaporation and applied for the adsorptive removal of Cu(II), and oxytetracycline (OTC) from wastewater.[76] The results showed that a hybrid CMC/GO/PANI thin film had a higher adsorption efficiency than pure materials due to the multifunctional synergetic effect. Hybrid CMC/GO/PANI thin film is a reusable adsorbent for eliminating pharmaceuticals and metal ions from wastewater, with a performance comparable to other previously reported systems, but unfortunately the cost of the GO and PANI is quite high.

Other recent examples consist of a novel eco-sustainable chitosan/gelatine/GO hybrid as an adsorbent for water purification from fluoroquinolonic antibiotics ofloxacin and ciprofloxacin as well as Pb.[77] The aerogel presents an antimicrobial effects after 1 h of incubation when employed as a filter for drinking water. Finally, efficient beads adsorbers for diclofenac (DIF) were obtained combining activated carbon (CAC) with chitosan in a 1 : 1 ratio. Notably, the active beads were obtained by switching the pH to basic conditions and the maximum adsorption capacity was 99.29 mg g^{-1}. Moreover, the hybrid materials were able to reduce DIF toxicity toward *Vigna mungo* seeds during germination. In fact, after filtration and DIF absorption the seed's growth was considerably improved.

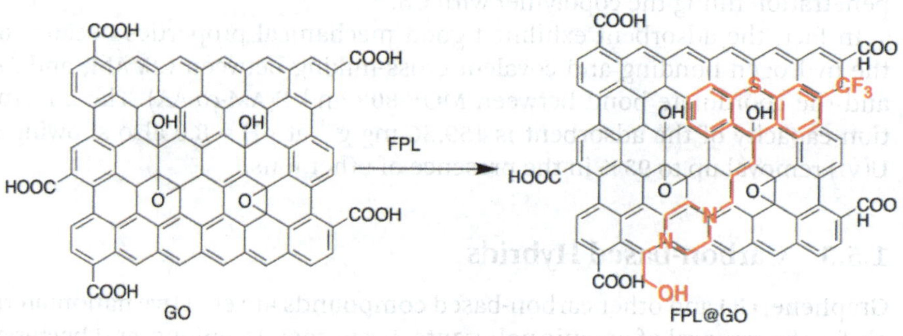

Figure 1.16 Representative absorption model of flupentixol (FPL) on GO*.[75]

1.6 Conclusions

The great variety of synthetic tools available to organic chemists to produce new functional materials or to add new functions to existing materials makes the possibility to design advanced materials for pollutant removal practically unlimited. However, the more complex is the need then the more expensive is the solution in terms of production costs of the advanced material, especially if it is needed in large quantities. In this context, the feasibility of a sustainable approach to water remediation by using advanced materials for pollutant removal is based upon a careful choice of the most appropriate technology/material as a function of the qualitative/quantitative composition of the water that needs to be treated. The suitability of a material for a given purpose, especially when removal is mainly based on adsorption processes, depends on the specificity of the interactions which, in turn, depend on the chemical moieties of the adsorbent framework and of the pollutant. Given the fact that the chemical features of the pollutant cannot be "designed", but rather their classification depends on the pollution source, it is important to be able to design the features of the adsorbent to optimize the response to the removal need. The choice of the best technology to implement as pollutant removal, will then rely on an accurate costs-versus-benefits analysis, including not only the efficiency of the removal, but also the costs of advanced material production and the availability of the source of the starting raw material. Once the efficacy of a given material is established, further developments should be focused on the possibility to re-use, regenerate, or recycle the used adsorbent, also through immobilization strategies for batch reactors or innovative separation approaches such as magnetic separation of heterogeneous components. In this framework, without the intention of providing exhaustive examples, this chapter intends to offer to the reader an overview of strategies available to obtain advanced materials, including hybrid composite materials, with features that can be combined as a function of the available primary source of the raw material (secondary waste product, biomasses, *etc.*), of the functionalization method, and of the type of target class of pollutants.

Acknowledgments

Carla Rizzo thanks PNR next generation EU-DM737/2021-CUP B79J21038330001 and FFR2022 for funding.

References

1. FAO, Clean water and sanitation, https://www.fao.org/family-farming/detail/en/c/1634522/.
2. D. Savoca, R. Melfi, A. Palumbo Piccionello, S. Barreca, S. Buscemi, V. Arizza, M. Arculeo and A. Pace, *Sci. Rep.*, 2021, **11**, 18763.

3. O. Al-Hashimi, K. Hashim, E. Loffill, T. Marolt Čebašek, I. Nakouti, A. A. H. Faisal and N. Al-Ansari, *Molecules*, 2021, **26**, 5913.
4. S. Sharma and A. Bhattacharya, *Appl. Water Sci.*, 2017, **7**, 1043–1067.
5. S. D. Hemming, J. M. Purkis, P. E. Warwick and A. B. Cundy, *Environ. Sci.: Processes Impacts*, 2023, **25**, 1909–1925.
6. J. Arun, N. Nirmala, P. Priyadharsini, S. S. Dawn, A. Santhosh, K. P. Gopinath and M. Govarthanan, *Mater. Lett.*, 2022, **310**, 131476.
7. S. A. Ashter, in *Introduction to Bioplastics Engineering*, ed. S. A. Ashter, William Andrew Publishing, Oxford, 2016, pp. 19–30. DOI: 10.1016/B978-0-323-39396-6.00002-6.
8. A. George, M. R. Sanjay, R. Srisuk, J. Parameswaranpillai and S. Siengchin, *Int. J. Biol. Macromol.*, 2020, **154**, 329–338.
9. P. R. Yaashikaa, P. Senthil Kumar and S. Karishma, *Environ. Res.*, 2022, **212**, 113114.
10. B. P. Charitha and P. Rao, *Carbohydr. Polym.*, 2017, **168**, 337–345.
11. M. F. Moradali and B. H. A. Rehm, *Nat. Rev. Microbiol.*, 2020, **18**, 195–210.
12. K. K. Sadasivuni, P. Saha, J. Adhikari, K. Deshmukh, M. B. Ahamed and J.-J. Cabibihan, *Polym. Compos.*, 2020, **41**, 32–59.
13. G. P. Udayakumar, S. Muthusamy, B. Selvaganesh, N. Sivarajasekar, K. Rambabu, S. Sivamani, N. Sivakumar, J. P. Maran and A. Hosseini-Bandegharaei, *Biotechnol. Adv.*, 2021, **52**, 107815.
14. C. Rizzo, G. Misia, S. Marullo, F. Billeci and F. D'Anna, *Green Chem.*, 2022, **24**, 1318–1334.
15. M. Stanisz, Ł. Klapiszewski and T. Jesionowski, *Chem. Eng. J.*, 2020, **397**, 125409.
16. M. R. Caruso, G. D'Agostino, S. Milioto, G. Cavallaro and G. Lazzara, *J. Mater. Sci.*, 2023, **58**, 12954–12975.
17. F. Qin, W. Yao, Y. Liu, B. Zhu, Q. Yang and Y. Zheng, *Cellulose*, 2023, **30**, 3653–3666.
18. Z. Chen, H.-N. Xu and X.-K. Ouyang, *Langmuir*, 2022, **38**, 5996–6003.
19. Z. Chen, Z. Huang, J. Liu, E. Wu, Q. Zheng and L. Cui, *J. Hazard. Mater.*, 2021, **406**, 124697.
20. J. Sun, Y. Wang, Y. He, J. Liu, L. Xu, Z. Zeng, Y. Song, J. Qiu, Z. Huang and L. Cui, *Carbohydr. Polym.*, 2022, **298**, 120059.
21. M. A. Bajaber, M. N. Anjum, M. Ibrahim, T. Farooq, M. N. Ahmad and Z. u. Abideen, *Molecules*, 2022, **27**, 8238.
22. G. B. D'Souza, A. Kumar, S. V. Kamath, A. S. Maraddi and S. K. Nataraj, *Chem. Eng. J.*, 2022, **443**, 136462.
23. M. Rigoletto, P. Calza, E. Gaggero and E. Laurenti, *Chem. Eng. J. Adv.*, 2022, **10**, 100252.
24. Y. Fei, Y. Li, S. Han and J. Ma, *J. Colloid Interface Sci.*, 2016, **484**, 196–204.
25. Y. Liu, C. Jin, Z. Yang, G. Wu, G. Liu and Z. Kong, *Int. J. Biol. Macromol.*, 2021, **187**, 880–891.
26. D. Meister, D. Ure, A. Awada, J.-C. Barrette, J. Gagnon, B. Mutus and J. F. Trant, *ACS Sustainable Chem. Eng.*, 2019, **7**, 20139–20150.

27. P. Deria, J. E. Mondloch, O. Karagiaridi, W. Bury, J. T. Hupp and O. K. Farha, *Chem. Soc. Rev.*, 2014, **43**, 5896–5912.
28. H. Furukawa, K. E. Cordova, M. O'Keeffe and O. M. Yaghi, *Science*, 2013, **341**, 1230444.
29. X. Zhang, Z. Chen, X. Liu, S. L. Hanna, X. Wang, R. Taheri-Ledari, A. Maleki, P. Li and O. K. Farha, *Chem. Soc. Rev.*, 2020, **49**, 7406–7427.
30. C. S. Diercks and O. M. Yaghi, *Science*, 2017, **355**, eaal1585.
31. M. Mon, R. Bruno, J. Ferrando-Soria, D. Armentano and E. Pardo, *J. Mater. Chem. A*, 2018, **6**, 4912–4947.
32. J. L. Obeso, J. G. Flores, C. V. Flores, M. T. Huxley, J. A. de los Reyes, R. A. Peralta, I. A. Ibarra and C. Leyva, *Chem. Commun.*, 2023, **59**, 10226–10242.
33. E. M. Dias and C. Petit, *J. Mater. Chem. A*, 2015, **3**, 22484–22506.
34. P. A. Kobielska, A. J. Howarth, O. K. Farha and S. Nayak, *Coord. Chem. Rev.*, 2018, **358**, 92–107.
35. S. M. Cohen, *J. Am. Chem. Soc.*, 2017, **139**, 2855–2863.
36. T. Grancha, J. Ferrando-Soria, H.-C. Zhou, J. Gascon, B. Seoane, J. Pasán, O. Fabelo, M. Julve and E. Pardo, *Angew. Chem., Int. Ed.*, 2015, **54**, 6521–6525.
37. D. M. Maklavany, Z. Rouzitalab, M. Bazmi, M. Askarieh and A. Nabavi-Pelesaraei, *ACS Sustainable Chem. Eng.*, 2023, **11**, 9816–9832.
38. C. Vogelsang, M. Grung, T. G. Jantsch, K. E. Tollefsen and H. Liltved, *Water Res.*, 2006, **40**, 3559–3570.
39. T. Wang, Z. Kou, S. Mu, J. Liu, D. He, I. S. Amiinu, W. Meng, K. Zhou, Z. Luo, S. Chaemchuen and F. Verpoort, *Adv. Funct. Mater.*, 2018, **28**, 1705048.
40. S. Y. Al nami and A. Hossan, *J. Mol. Liq.*, 2023, **382**, 122065.
41. L. Wang, T. Liu, J. Cheng, H. Zou, J. Lu, H. Liu, Y. Li, J. Dou and S. Wang, *J. Mol. Struct.*, 2024, **1296**, 136815.
42. L. Rasuli, M. H. Dehghani, M. Alimohammadi, K. Yaghmaeian, N. Rastkari and M. Salari, *J. Mol. Liq.*, 2021, **339**, 116801.
43. R. Li, X. Tang, J. Wu, K. Zhang, Q. Zhang, J. Wang, J. Zheng, S. Zheng, J. Fan, W. Zhang, X. Li and S. Cai, *Chem. Eng. J.*, 2023, **464**, 142706.
44. B. Azhar, C. Avian and A. H. Tiwikrama, *J. Mol. Liq.*, 2023, **389**, 122844.
45. A. Sharma, M. Kumari, M. Tahir, S. Jain, S. Sharma and N. Kumar, *J. Mol. Liq.*, 2023, **386**, 122429.
46. C. Fonseka, S. Ryu, Y. Choo, M. Mullett, R. Thiruvenkatachari, G. Naidu and S. Vigneswaran, *ACS Sustainable Chem. Eng.*, 2021, **9**, 16896–16904.
47. X. Liu, F. Gao, T. Jin, K. Ma, H. Shi, M. Wang, Y. Gao, W. Xue, J. Zhao, S. Xiao, Y. Ouyang and G. Ye, *Nat. Commun.*, 2023, **14**, 5097.
48. A. Wang, J. Ni, W. Wang, D. Liu, Q. Zhu, B. Xue, C.-C. Chang, J. Ma and Y. Zhao, *Appl. Catal., B*, 2022, **319**, 121926.
49. A. Hassan, M. M. R. Mollah, S. Das and N. Das, *J. Mater. Chem. A*, 2023, **11**, 17226–17236.
50. M. Mon, R. Bruno, E. Tiburcio, M. Viciano-Chumillas, L. H. G. Kalinke, J. Ferrando-Soria, D. Armentano and E. Pardo, *J. Am. Chem. Soc.*, 2019, **141**, 13601–13609.

51. A. Maio, R. Scaffaro, L. Lentini, A. Palumbo Piccionello and I. Pibiri, *Chem. Eng. J.*, 2018, **334**, 54–65.
52. C. Santhosh, V. Velmurugan, G. Jacob, S. K. Jeong, A. N. Grace and A. Bhatnagar, *Chem. Eng. J.*, 2016, **306**, 1116–1137.
53. S. Álvarez-Torrellas, J. A. Peres, V. Gil-Álvarez, G. Ovejero and J. García, *Chem. Eng. J.*, 2017, **320**, 319–329.
54. N. Baig, I. Ihsanulla, M. Sajid and T. A. Saleh, *J. Environ. Manage.*, 2019, **244**, 370–382.
55. Y. Ngernyen, D. Petsri, K. Sribanthao, K. Kongpennit, P. Pinijnam, R. Pedsakul and A. J. Hunt, *RSC Adv.*, 2023, **13**, 14712–14728.
56. T. J. M. Fraga, M. N. Carvalho, M. G. Ghislandi and M. A. da Motta Sobrinho, *Braz. J. Chem. Eng.*, 2019, **36**, 1–31.
57. S. Verma, K.-H. Kim, N. Kumar, S. S. Bhattacharya, M. Naushad and R. K. Dutta, *J. Hazard. Mater.*, 2022, **429**, 128308.
58. K. Thakur and B. Kandasubramanian, *J. Chem. Eng. Data*, 2019, **64**, 833–867.
59. M. Sajid, I. Ihsanullah, M. Tariq Khan and N. Baig, *Sep. Purif. Technol.*, 2023, **305**, 122453.
60. C. Sun, Z. Wang, L. Chen and F. Li, *Chem. Eng. J.*, 2020, **393**, 124796.
61. A. Balati, M. Ghanbari, K. S. Behzad and M. M. Amini, *Acta Chim. Slov.*, 2017, **64**, 479–490.
62. H. Basu, S. Saha, S. K. Kailasa and R. K. Singhal, *Environ. Sci.: Water Res. Technol.*, 2020, **6**, 3214–3248.
63. K. Nava-Andrade, G. G. Carbajal-Arízaga, S. Obregón and V. Rodríguez-González, *J. Environ. Manage.*, 2021, **288**, 112399.
64. S. Nayab, H. Baig, A. Ghaffar, E. Tuncel, Z. Oluz, H. Duran and B. Yameen, *RSC Adv.*, 2018, **8**, 23963–23972.
65. T. S. Machado, L. Crestani, G. Marchezi, F. Melara, J. R. de Mello, G. L. Dotto and J. S. Piccin, *Carbohydr. Polym.*, 2022, **277**, 118868.
66. M. H. Mruthunjayappa, V. T. Sharma, K. Dharmalingam, N. Sanna Kotrappanavar and D. Mondal, *ACS Appl. Bio Mater.*, 2020, **3**, 5233–5243.
67. S. Rojas and P. Horcajada, *Chem. Rev.*, 2020, **120**, 8378–8415.
68. F. Zadehahmadi, N. T. Eden, H. Mahdavi, K. Konstas, J. I. Mardel, M. Shaibani, P. C. Banerjee and M. R. Hill, *Environ. Sci.: Water Res. Technol.*, 2023, **9**, 1305–1330.
69. M. Allahbakhshi, N. M. Mahmoodi, M. Mosaferi, H. Kazemian and H. Aslani, *Surf. Interfaces*, 2022, **35**, 102471.
70. A. A. Ghani, K. C. Devarayapalli, B. Kim, Y. Lim, G. Kim, J. Jang and D. S. Lee, *Carbohydr. Polym.*, 2023, **318**, 121098.
71. Y. Song, H. Li, T. Shan, P. Yang, S. Li, Z. Liu, C. Liu and C. Shen, *Carbohydr. Polym.*, 2023, **302**, 120377.
72. C. Li, Q. Xu, S. Xu, X. Zhang, X. Hou and P. Wu, *RSC Adv.*, 2017, **7**, 16204–16209.
73. S. Dayana Priyadharshini, S. Manikandan, R. Kiruthiga, U. Rednam, P. S. Babu, R. Subbaiya, N. Karmegam, W. Kim and M. Govarthanan, *Environ. Pollut.*, 2022, **306**, 119377.

74. C. Rizzo, S. Marullo and F. D'Anna, *Environ. Sci.: Nano*, 2021, **8**, 131–145.
75. K. Balasubramani, N. Sivarajasekar, S. Muthusaravanan, K. Ram, M. Naushad, T. Ahamad and G. Sharma, *J. Mol. Liq.*, 2020, **319**, 114371.
76. M. A. Taleb, R. Kumar and M. A. Barakat, *Int. J. Biol. Macromol.*, 2023, **253**, 126699.
77. A. Kovtun, E. Campodoni, L. Favaretto, M. Zambianchi, A. Salatino, S. Amalfitano, M. L. Navacchia, B. Casentini, V. Palermo, M. Sandri and M. Melucci, *Chemosphere*, 2020, **259**, 127501.

CHAPTER 2

Synthesis and Functionalization of Advanced Materials for Pollutant Removal

KALAIVIZHI RAJAPPAN[a*], NEERAJA BOSE[a], ANISHIA AMBROSE[a] AND ANWAR IQBAL[b]

[a]Department of Chemistry, Faculty of Engineering and Technology, SRM Institute of Science and Technology, Kattankulathur, Tamilnadu, 603203, India; [b]School of Chemical Sciences, Universiti Sains Malaysia, 11800 Gelugor, Penang, Malaysia
E-mail: kalaivir@srmist.edu.in

2.1 Introduction

In aquatic ecosystems, the impurities present beyond the permissible limits are referred to as water pollution. Due to rapid progress in modernization and suburbanization, uncontrolled wastewater discharge into the water ecosystem is the root cause of water contamination. The outflow of untreated water effluents from various industries, such as agricultural industries, mining industries, pharmaceutical industries, electrical product manufacturing industries, *etc.*, are sources of water pollution.[1] Pollutants such as inorganic (heavy metals) and organic pollutants (organic dyes, aromatic compounds, polynuclear aromatic hydrocarbons) are responsible for water pollution.[2] To overcome such difficulties, removing such contaminants through advanced functionalized materials is

Chemistry in the Environment Series No. 14
Advanced Materials for Emerging Water Pollutant Removal
Edited by Pei Sean Goh, Devagi Kanakaraju, Anwar Iqbal and Ahmad Fauzi Ismail

Published by the Royal Society of Chemistry, www.rsc.org

crucial to improving the water ecosystem. Nanomaterials are the emerging advanced materials for removing such pollutants from water systems. Materials that are reduced to the nanoscale experience a higher surface-to-volume ratio than the bulk material. They have high regeneration abilities and reusability.[3] Metal oxides, bimetallic compounds, magnetite nanoparticles, micelles, dendrimers, nanotubes, nano clays, graphene oxide (GO), covalent organic frameworks (COF), metal organic frameworks (MOF),[4] polymeric nanoparticles, and MXenes[5] are examples of the emerging nanomaterials used for the removal of contaminants from water bodies.

The MOFs and MXenes are found to be the most suitable and potent materials for the removal of impurities from aquatic ecosystems. The MOFs are coordination compounds composed of metal ligands and organic linkers. They may have 2D or 3D structures according to the structure of the organic linker. They have unique properties such as tunable porosity and band gap, high selectivity and crystallinity, and modifiable structures that are very important for removing contaminants.[6] MXenes are emerging 2D materials composed of transition metal (d-block elements) and their carbides, nitrides, or carbonitrides. These materials have properties such as high sorption selectivity, hydrophilicity, high metallic conductivity, and they also activate the metallic hydroxide sites. Due to the existence of several functional groups such as –O– and –OH, as well as an increased surface area, MXene not only provides the sites for ion exchange but also possesses the capability to reduce certain positively charged ions and organic molecules. The combination of reduction capability with adsorption properties makes MXenes a very efficient material for pollutant removal.[7]

Biomass has been converted to biosorbent composites, polymer composites, activated charcoal composites and hybrid composites, and functionalized with various chemical compounds to increase their efficiency in removing pollutants from water.[8] Organic functionalization, amine functionalization, catechol functionalization, grafting, dipping, coating, and inorganic material-based modification are among the functionalization techniques often used. For instance, increasing the active sites on the surface of magnetic nanoparticles by surface functionalizing with inorganic substances, molecules of organic matter, polymers made from organic materials, natural biopolymers, synthetic polymers, and synthetic natural composite polymers. Additionally, it improves performance, lowers coagulation, and promotes stability.[9] MXenes with surface-modified surfaces exhibit improved antibacterial, radioactive, and organic dye ion removal capabilities.

This chapter aims to highlight the various techniques of synthesizing and functionalization of the advanced materials with respect to changes in morphology and surface area. In conclusion, the challenges and future perspectives of the materials are also discussed.

2.2 Advanced Materials for Pollutant Removal

2.2.1 Covalent Organic Framework

Covalent organic frameworks (COF), which are crystal porous materials, have attracted a lot of attention in the field of pollutant treatment due to their well-ordered nanopores and amenable pore environment for functionalization.[10] COFs can be varied in their structure and stability by the source of building monomers such as knots, edges, linkages, imine, hydrazone, and azine. Both polar and non-polar organic solvents can stabilize the pore structure of COFs. Since they can be employed as membrane materials in both aqueous and organic solvents, COFs are seen as attractive candidates.[11] COFs stand out from conventional sorbents because of their notable advantages such as tuneable and organized structure, high specific surface area, higher physical and chemical stability, high adsorption capacity,[12] and customizable shape. Outstanding benefits of COFs include adaptable structure and high chemical and physical stability levels.[13] Their potent covalent bonds and affinity for interaction with material *via* π–π interactions made COFs more stable structures for removing pollutants from another sorbent (Figure 2.1).

2.2.2 Metal Organic Framework

Metal organic frameworks (MOF) are infinite crystal structures made by clusters of metals bonded to organic ligands through coordination; they are very porous and adaptable due to their tunability in structure and chemicals. Due to their vast surface areas, stable porosity, varied structural makeup, and adaptable functional groups, MOFs have substantial advantages in several applications, including storage, separation (primarily used as an adsorbent material for separations of gas and liquid[16]), sensing, catalysis,

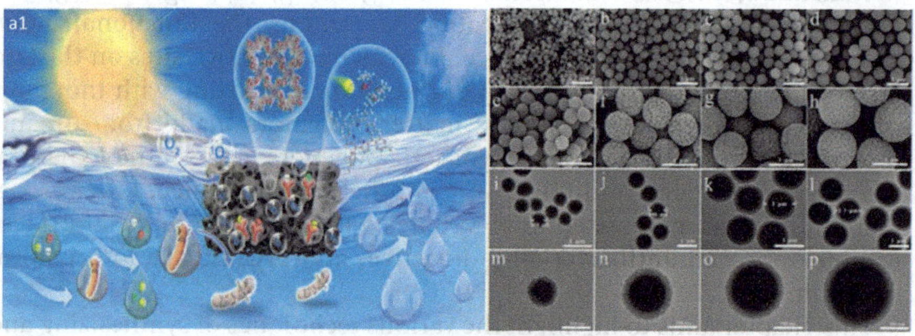

Figure 2.1 (a1) COF for water treatment. Reproduced from ref. 14 with permission from Elsevier, Copyright 2022. (a–p) SEM representation of size controllable spherical COF. Reproduced from ref. 15 with permission from American Chemical Society, Copyright 2019.

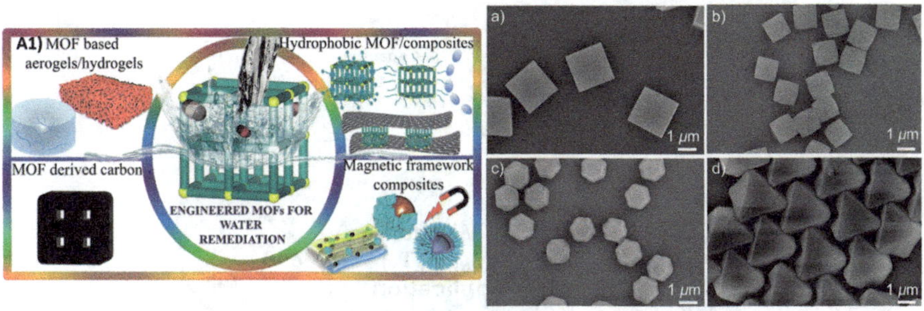

Figure 2.2 (A1) MOF for water remediation. Reproduced from ref. 19 with permission from Elsevier, Copyright 2021. (a–d) MOF monodispersed size tunable crystals. Reproduced from ref. 20 with permission from Elsevier, Copyright 2018.

bioimaging, and other fields.[17] MOFs are potential materials as photosensitive compounds which have optical characteristics and high photocatalytic properties resulting in high efficiency for light-sensitized removal and photosensitized removal of pollutants. Regardless, MOFs can be very unstable in water depending on the type of binding between the metallic sites and the organic ligands. A sustainable and novel combination of MOF which can show a stable nature in water is today's research concern. These materials can tolerate prolonged submersion in water over a wide pH range (0–12). Due to their high porosities and adsorbate–adsorbent interactions, MOFs can generally be seen as viable options for the treatment of wastewater streams. They may also convert chemicals into H_2 and employ their catalytic properties to degrade hazardous molecules[18] (Figure 2.2).

2.2.3 MXene

MXene is a developing novel material that includes 2D particles derived from graphene and expanded to metal oxides (MOs), 2D polymer matrixes, transition metal-based dichalcogenides, *etc.* MXene, has the formula $M_{n+1}X_nT_x$, a group of 2D transition metal-based carbides and nitrides that are clogged on the surface of metals and have an imperfect spread of different functional groups. The M stands for the transition metal, X for carbon(C) or nitrogen(N), T for functional groups like hydroxyl(–OH), oxygen(–O–), or fluorine(–F), and *n* is an integer.[21] MXene was produced to address the shortcomings of conventional components.

MXenes with –O–, –OH, and –F terminations offer outstanding water permeance, good filtration ability, and extended operational life for water purification.[5] MXene has extreme electrical conductivity (2.105 S m^{-1}) and is comparable with multilayered graphene molecules. It has increased specific surface area in comparison, improved surface mechanism due to its functionalization capability, and easy dispersion in water. The 2D polymer-based MXene can efficiently remove phosphate and nitrate ions from

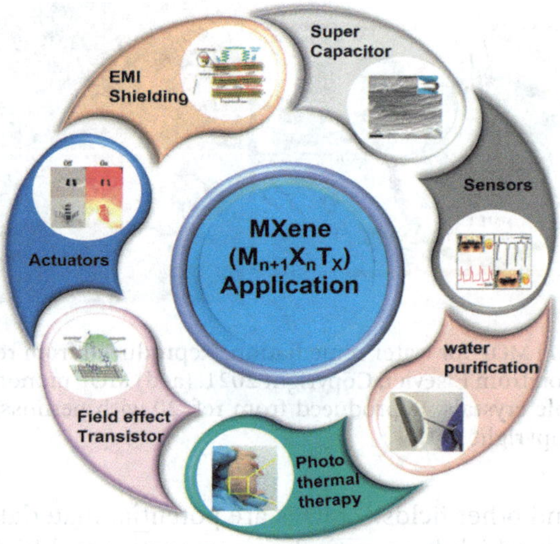

Figure 2.3 Pictorial representation of various applications of Mxene. Reproduced from ref. 22, https://doi.org/10.1002/gch2.202100120, under the terms of the CC BY 4.0 license https://creativecommons.org/licenses/by/4.0/.

aquatic environments. This is due to the surface affinity of the negatively charged anions on the surface of MXenes, resulting in surface complexation and electrostatic attraction with phosphate and nitrate ions. Exfoliation of an "A" layer from a "MAX" precursor can be used to create these materials; M, A, and X stand for transition metals (like Ti), group IIIA or IVA elements (like Ti), and N or C, respectively. MXene can be incorporated with polymeric membranes, capacitive deionization, photocatalysis, electrochemical separation, and adsorbents for water treatment and purification[7] (Figure 2.3).

2.2.4 Bio-based Materials

A material rich in carbon is produced by heating biomass in an environment with little to no oxygen, known as biochar. Biochar is a perfect adsorbent for extracting pollutants from aqueous solutions due to its enormous specific surface area, porous structure, availability of surface functional groups, and mineral constituents. Biochar can eliminate toxins and carcinogenic chemicals, including medications, metallic and non-metallic pollutants, colour, taste, and odour from aqueous solutions.[23]

Conversely, "biocomposite" is a composite material comprising a matrix (resin) and reinforcement of natural adsorbent. The rising environmental crisis and high cost of synthetic adsorbents established the use of biocomposites in the field of pollutant removal.[24] Various natural biopolymers such as agricultural biocomposites, alginates, chitin, silk fibroin, and cellulose

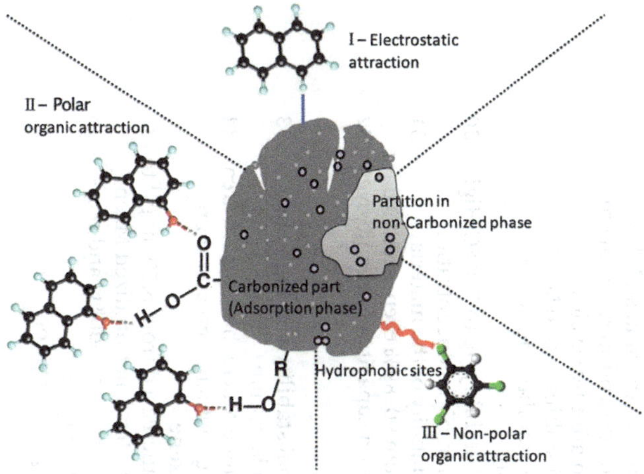

Figure 2.4 Interaction mechanism of biochar with organic pollutants. Reproduced from ref. 26 with permission from Elsevier, Copyright 2014.

and its derivatives are widely used biocomposites in the field of pollutant removal from water.[25] The adsorption mechanism between biochar and pollutants would depend on the electrostatic attraction of pollutants onto the hydrophobic site of biochar. Also, the organic pollutant tends to be adsorbed on the carbonized part of prepared biochar. A small phase of biochar can be non-carbonized where there less adsorption of pollutants takes place. A pictorial representation of the removal mechanism of biochar is represented in Figure 2.4.

2.3 Functionalization of Advanced Materials

Table 2.1 shows the improvement in the physiochemical properties of the advanced materials by functionalizing using various techniques.

2.3.1 Functionalized COF for Pollutant Removal

COF was functionalized by sulfonic acid for the elimination of pharmaceutical and individual care products from water and wastewater. The sulfonic acid ($R{-}S({=}O)_2{-}OH$) functionalized COF was obtained *via* the solvothermal technique. They seem to produce a greater affinity for examining 13 drugs present in water sources. As a result, the functionalized COF had prolonged flocculent growth and a stably maintained sphere due to the bridging of 1,2-ethanedisulfonic acid. The sulfonic acid functionalized COF showed relatively higher adsorption than unmodified COF due to the greater probability of π–π interaction. The hydrogen bonding and the active adsorption sites on the surface of the material could explain the adsorption mechanism. The oxygen atom in $-SO_3$ and –H of the –NH

Table 2.1 Functionalization of advanced materials by various techniques and their improved properties.

Materials	Functionalization	Technique	Morphology	Properties	References
COF	Thiol functionalized magnetic COF	Grinding	Stacking interaction of sheets or platelet-like polymer crystallites (porous material)	High thermal stability up to 300 °C Higher Hg^{2+} removal than pristine COF	27
	Sulfonic group functionalized COF	COF linkages	Porous material	High chemical and thermal stability Maintained adsorption removal efficiency even after five cycles	28
	Benzimidazole functionalized 2D COF	COF linkages and template	Small spherical particles	High thermal stability up to 500 °C Irradiation stability and structural adjustability	29
MOF	GO@ZIF-8	Imidazole linking	Hexagonal	High thermal stability up to 750 °C High adsorption in high concentration	30
	Amino@Magnetic MOF	Linking method	Not mentioned	The storage stability of the material retained from 90–89% even after 29 days	31
	PPI-dendrimer functionalized magnetic MOF	Hydro thermal and Sonication	Semi-spherical	High adsorption capacity of organic dyes	32
Mxene	Amino acid functionalized bioinspired Mxene	Self-polymerization	Packed stacked lamellar layered	1070 K thermal stability and high efficiency of adsorption	33
	Biosurfactant-Mxene	Etching and sonication	Nanosheets	Surface modified Mxene has surface area of 22.5 $m^2\ g^{-1}$ which is greater than pristine Mxene	34
	PEI-Mxene	Self-assembly	Single layered nanosheets	Surface area increased from 5.59 $m^2\ g^{-1}$ to 32.39 $m^2\ g^{-1}$ for functionalized Mxene	35
Biochar and biocomposite	Nitrogen functionalized Biochar	One pot synthesis	Microparticle with porous feature	Thermal stability up to 500 °C and recyclable for four cycles	36
	Acid functionalized Biochar	Pyrolysis	—	511 $m^2\ g^{-1}$, high adsorption	37
	Amino functionalized Biochar	Hydrothermal	Irregular morphology with distinct porous nature	91.8 $m^2\ g^{-1}$ of surface area and had regenerative ability of 80% of Cr(VI) of 86%, Sb(V) of 91%	38
	Amine functionalized rice husk magnetic biocomposite	One step solvothermal	—	19.95 $m^2\ g^{-1}$, thermal stability up to 800 °C	39
	Chitosan/PVA/AC biocomposite	Dispersion	Smooth and porous spherical bead like structure	Increased adsorption in presence of pH, concentration and temperature	40

in DCF (dichlorofluorescein) will ionize by creating an intermolecular hydrogen bond. Additionally, the relationship between porous adsorbents and DCF is significantly influenced by hydrogen bonding.[41] The obtained COF is recyclable *via* a simple water and methanol rising method. Thus, it was concluded that this COF was a potential material for the treatment of pharmaceutical and personal care products from wastewater. Thiol-based COF was prepared for the removal of mercury from water *via* salt-mediated crystallization and subsequent cutting strategy. Thiol functionalized COF has inherited the advantages of COF as well as showing magnetic properties. Furthermore, the salt-mediated method produces a high crystalline and ultra-porous structure. Observing its morphological analysis, the prepared COF represents a perfect porous nature with stacking interactions between the platelet-like polymer crystallites or its sheets. The mercury removal efficiency from the wastewater by the prepared thiol functionalized COF shows a rate of 99%. This high rate results from the soft–soft interaction between the S atom of thiol functionalized COF and mercury in water sources.[42]

2.3.2 Functionalization of MOF for Pollutant Removal

APTMS (3-aminopropyl-trimethoxysilane) functionalized ZIF-8 MOF framework was prepared for the removal of Direct Red 23(DR23). The preparation of APTMS functionalized ZIF-8 was based on a simple and viable stirring process. The prepared MOF material has achieved a perfectly regular and uniform hexagonal structure. The functionalized ZIF-8 material showed a surface area of 1128.20 $m^2 g^{-1}$, whereas the unmodified ZIF-8 showed a surface area of 1269.01 $m^2 g^{-1}$. The APTMS functionalized ZIF-8 material was also examined with its antibacterial activity, which showed 90% of bacterial death with Gram-negative bacteria *S. aureus*. The APTMS functionalized ZIF-8 shows a removal efficiency of more than 95% at pH-2.20 whereas pure ZIF-8 showed less than 20% removal efficiency. The regenerative efficiency of the material seems to be repeatedly achieving more than 95% removal efficiency for three consecutive periods.[43] CNT/Fe_2O_3/ZIF-8 and NH_2-CNT/Fe_2O_3/ZIF-8 were synthesized through an ultrasound-assisted technique. The ZIF-8 showed a dodecahedron structure and the NH_2-CNT/Fe_2O_3 compounds were better connected by the improved ZIF-8. The surface area of the prepared functionalized material is 1659 $m^2 g^{-1}$ whereas pure ZIF-8 has a lesser surface area of 1485 $m^2 g^{-1}$. The functionalized as well as pristine ZIF-8 were examined for selective removal of cationic dyes Malachite Green (MG) and Rhodamine B (Rh B) from the binary system. The method used for the removal of MG and RhB is adsorption. The optimum adsorption capacity is found that 25 mg L^{-1} dye solution is adsorbed within 145 min at pH 6 and at room temperature. The amount of adsorbent used is only 0.004 g. The dye removal efficiency was optimized using response surface methodology (RSM) artificial neural networks.[44]

2.3.3 Functionalization of MXene

Highly stable MXene functionalized with APTES and chitosan were fabricated for the adsorption of chromium from the industrial polluted water. Firstly, the authors etched the commercially available Ti_3C_2 powder in a polytetraethylene reaction kettle (etching is the process of removing aluminium from the commercial Ti_3AlC_2 arrangement). Secondly, after etching, the obtained material was functionalized by APTES on stirring with TEA (triethyleneamine) and by the slow addition of APTES (3-aminopropyl)triethoxysilane. The etched material functionalization of APTES was done with stirring. As, a result the functionalized MXene was reported to have a smooth and integrated structure. Polyimidazole-modified MXene (Ti_3C_2@IMIZ) was also produced and utilized for the adsorption of Cr(VI) from sewage treatment water. The adsorption process was observed to be pH-dependent. When pH increases the adsorption capacity decreases whereas the maximum adsorption is achieved at pH 2 with the removal efficiency of 119.5 mg g^{-1}. The pH of the medium affects the surface charge and bonding sites of adsorbents leading to efficient adsorption of heavy metal ions from the solution. On the other hand, due to the presence of free amino groups on the adsorbent surface some of the Cr(VI) ions were partially reduced to Cr(III) from stable complexes resulting in decrease in the removal efficiency. Overall, the APTES-MXene and IMIZ-MXene were more efficient than those of pristine MXene.[45]

Carboxyl functionalized MXene was prepared for the removal of U(VI) and Eu(III) by functionalizing commercially available Ti_3C_2 with phenylcarboxylic acid diazonium salt. The TEM analysis indicates a nanoflake-like morphology. The carboxyl functionalized MXene has shown 80% radionuclide removal from artificially generated complicated water, after 3 h the material showed 90% removal efficiency. They concluded that the carboxyl functionalized MXene shows an adsorption mechanism through the carboxyl groups on the surface which forms a complex with the radionuclide ions. It has maximum removal efficiency for U (344.8 mg g^{-1}) and Eu (97.1 mg g^{-1}). Thus, carboxyl functionalized MXene has the potential to adsorb complex ions from the water sources.[46]

2.3.4 Functionalization of Biochar and Biocomposites

Biochar from rice straw was prepared using an electric tube furnace in an oxygen-limited atmosphere by supplying nitrogen gas for 30 min at a temperature of 600 °C. Also, double-layered hydroxide (LDH) functionalized biochar was prepared using $AlCl_3\cdot 6H_2O$ and $MgCl_2\cdot H_2O$ as precursors at pH 10 using 1 M NaOH and incubated at 80 °C at different compositions of biochar. The functionalized biochar has a morphology of a smooth and tubular structure, which is different from pristine biochar. The author claimed that the functionalization of LDH on biochar was uniform on the surface and also present in the pore channels. Swapping of ions, diffusion of pores, electrostatic attraction, and precipitation are the major prevalent mechanisms for the adsorption of pollutants using pure and functionalized biochar. The adsorption of phosphate was higher at lower pH. At lower pH, a considerable positive charge built up on the adsorbents' surfaces, facilitating strong electrostatic interactions with negatively charged phosphate species. The capability of phosphate–OH ion exchange was made possible by the hydroxyl groups present in the adsorbents. This behaviour can be induced by forming covalent chemical interactions between the phosphate ions and metal–oxide surfaces (Mg–O and Al–O) or surface-active sites in monodentate and bidentate inner-sphere surface complexations.[47]

Biocomposite was produced from carrageenan, lime, and lemon agro-waste biochar. The biochar is prepared by heating at 65 °C for 48 h, ground and sieved further to a size smaller than 0.250 mm. Furthermore, the biocomposite was prepared by enzymatic modification using two processes. The prepared biochar was incorporated with laccase through fixation and adsorption. The limitation associated with this kind of biocomposite is its heterogeneity. However, the adsorption capacity of the material was improved when converted to biocomposite.[48]

2.4 Synthesis Techniques for Advanced Functionalized Materials

There are several techniques which have been adopted for the synthesis of advanced functionalized materials. Depending on the synthesis technique, the removal as well as the adsorption efficiency also changes. Table 2.2 explains the various methods for the synthesis of advanced materials.

2.4.1 Solvothermal

COF, MOF, MXene, and various origins of biochar and biocomposites are synthesized *via* the solvothermal technique. This technique uses suitable solvents in the presence of high pressures and temperatures (1–10 000 atm, typically). To maintain a constant high pressure and temperature these procedures are carried out in a Teflon-lined sealed vessel (such as an autoclave) beyond the boiling point of the solvents. Under these conditions, due to the elevated pressure, the solvent remains in the liquid state beyond its boiling point and crystals with unique structures start to grow. Further, the crystalline porous advanced materials are retrieved from the autoclave after cooling to room temperature.[106] The solvothermal method is most commonly used in synthesizing MOF which can provide diverse morphology with common organic solvents like dimethylformamide (DMF), acetone, diethyl formamide, methanol, acetonitrile, and ethanol.[107] The unique advantage of synthesizing functionalized materials *via* solvothermal material is that the procuring material has higher stability in a novel composition. Thus, this method is useful in obtaining material with increased stability and excellent removal efficiency on reuse (Figure 2.5).

Table 2.2 Various methods for the synthesis of advanced materials.

Method	Material	Linkers	Morphology	Pollutants	Efficiency		References
					Adsorption	Removal	
Microwave assisted synthesis	COF	Terephthalaldehyde	Inverse spinel	Hg^{2+}	97.65 mg g^{-1}	—	49
		2,3,5,6-Tetrafluoro-4-pyridinecarbonitrile and 2,3,6,7,10,11-hexahydroxytriphenylene	Reticulate morphology	Perfluorinated alkyl substances	—	89.5%–105%	50
		$Cu(NO_3)_2{\cdot}6H_2O$, H_3btc and $H_3PW_{12}O_{40}$	Octahedral crystal	Pb^{2+}, Cd^{2+}	98.18 mg g^{-1}, 32.45 mg g^{-1}	—	51
	MOF	Fumaric acid	Hexagonal rod	Cd(II), Pb(II), Cr(VI)	693 mg g^{-1}, 536.22 mg g^{-1}, 7.99 mg g^{-1}	—	52
		2-Aminoterephthalic acid	Needle	Methylene blue	—	>80%	53
		Terephthalic acid	Octahedral morphology	Mixture of dyes	98%	—	54
		Terephthalic acid	Octahedral morphology	Acid chrome blue K	591.79 mg g^{-1}	—	55
	Mxene	Titanium powder, aluminum powder, graphite powder, tin powder	Accordion like structure	Other application	—	—	56
	Biochar	Rise husk, β-cyclodextrin, glutaraldehyde	Rough and lamellar	Bisphenol A, Pb(II)	209.20 mg g^{-1}, 240.13 mg g^{-1}	—	57
		Walnut shell, ferric chloride	Heterogeneous and porous	Salicylic acid, naproxen, ketoprofen	683 mg g^{-1}, 533 mg g^{-1}, and 444 mg g^{-1}	—	58
		Jackfruit peel and mushroom	Spherical	Nitrates and phosphates	5.26 mg g^{-1} (real sample) and 7.94 mg g^{-1} (lab scale)	—	59
	Biocomposite	Epichlorohydrin (EPI) and glutaraldehyde (GA)	Rough with white particles	Pb(II)	For (EPI) – 216.06 mg g^{-1}, and for (GA) – 279.08 mg g^{-1}	—	60
		$Fe(NO_3)_3{\cdot}9H_2O$, NaOH, $Zn(NO_3)_2{\cdot}6H_2O$, methylcellulose	Smooth, uniform, and compact	Metronidazole	—	92.65% (synthetic sample) and 71.12% (real sample)	61

Solvothermal synthesis	COF	TAB-DVA-COF, 2,2-azobisisobutyronitrile, 1,2-ethanedisulfonic acid	Spherical	Non-steroidal, sulfonamides, fluoroquinolones, and bisphenols	—	DCF (non-steroidal pollutant) (770 mg g^{-1})	41
		Tb, 3,3′-bis(trifluoromethyl)benzidine, 1,4-dioxane, mesitylene and acetic acid	Round shape	Ibuprofen	119 mg g^{-1}	—	62
	MOF	2,6-Naphthalenedicarboxylic acid	Rough	Rhodamine B, Rh6G	219.78 mg g^{-1} and 306.75 mg g^{-1}	—	63
		Terephthalic acid, zirconium(IV) chloride	—	Rhodamine 6G, propranolol, and 16 bisphenol A	—	96.0%	64
		Zirconyl chloride octahydrate, formic acid, DMF	—	Methylene blue, methyl orange	1239 mg g^{-1}, 1022 mg g^{-1}	—	65
	Mxene	Ti_3AlC_2, HCl, LiF	Accordion morphology	Methylene blue, rhodamine B	—	92.37%	66

(*continued*)

Table 2.2 (*Continued*)

Method	Material	Linkers	Morphology	Pollutants	Efficiency		References
					Adsorption	Removal	
Hydrothermal synthesis	COF	2,4,6-Tris(4-formyl phenoxy)-1,3,5-triazine and 2,4-diamino-6-phenyl-1,3,5-triazine	Needle-like crystalline structure	DFS-4BS	$8501\ mg\ g^{-1}$	—	67
		Benzoguanamine, trialdehyde, *n*-butanol/*o*-dichlorobenzene	Needle-like crystal	Hg^{2+}	$1826\ mg\ g^{-1}$	—	68
	MOF	2,5-Dihydroxyterephthalic acid	Agglomerated cubic shape	Methylene blue and tetracycline hydrochloride	99.96%, 94.88%	—	69
		$Cr(NO_3)_3{\cdot}9H_2O$, terephthalic acid	Octahedral	Acid green 25 and Reactive yellow 186	—	93%	70
		Zirconium chloride, europium nitrate, terephthalic acid, methylamine aqueous solution	Agglomerated crystalline	Tetracycline	$289\ mg\ g^{-1}$	—	71
	Mxene	Nb_2AlC, HF, tetramethyl ammonium hydroxide	Accordion-like multilayer	Rhodamine B, methylene blue, tetracycline hydrochloride	—	99.8%, 92.7%, 83.1%	72
		Ti_3AlC_2, HF	Accordion-like structure	Carbamazepine	—	98.67%	73

Template synthesis	COF	2,5-Dihydroxyterephthalaldehyde, 1,3,5-tris(4-aminophenyl)benzene or *N*1,*N*1-bis(4-aminophenyl)benzene-1,4-diamine in mesitylene-dioxane, acetic acid	Hollow spherical	Other application	—	—	74
	MOF	Copper acetate(II), benzene-1,3,5-tricarboxylate and zinc nitrate hexahydrate, 2-methylimidazole	Truncated crystals	Congo red	—	95%	75
		Cu $(NO_3)_2{\cdot}3H_2O$ and H_3BTC	Pyramid	Vanadium	285.71 mg g^{-1}	—	76
	Mxene	2D h-MoO_3@NaCl2D h-WO_3@KCl2D h-VOX@NaCl	Nanosheets	Other application	—	—	77
Linkers	COF	*p*-Toluene sulfonic acid (monohydrate), *p*-phenylenediamine, MNP-NH_2, Tp	—	Diclofenac sodium	565 mg g^{-1}	—	78
	Mxene	Ti_2AlC, HCl, LiF	Multilayered structure	U(VI)	>90%	—	79

(*continued*)

Table 2.2 (*Continued*)

Method	Material	Linkers	Morphology	Pollutants	Efficiency		References
					Adsorption	Removal	
Sonochemical synthesis	COF	*p*-Toluene sulphonic acid monohydrate, 1,3,5-triformylphloroglucinol, diamine *p*-azoaniline	Sheet like layers	Bisphenol-A	—	85.7%	80
	MOF	2,5-Thiophene dicarboxylic acid and *N*,*N*-bis-pyridin-4-yl-methylene-hydrazine	Nanoplates	2,4-Dichlorophenol, Amoxicillin	—	>90%	81
		Fumaric acid	Regular	Methyl Violet 2B	90%	—	82
		4,4-Oxybisbenzoic acid and 4,4-bipyridine	—	Hg(II)	338 mg g^{-1}	—	83
		N,*N*′-Bis-(4-pyridylformamide)-1,4-benzenediamine, *N*,*N*-dimethylformamide	Square shaped thin nanoplates	Methylene blue	90%	—	84
	Mxene	ML-Ti_3C_2Tx	Sheet	Cu^{2+}, Pb^{2+}, Cd^{2+}, and Cr^{3+}	—	98.29%, 90.04%, 52.90%, and 73.64%	85
Plasma	MOF	($Fe(NO_3)_3·9H_2O$), 1,3,5-benzenetricarboxylic acid, tetracycline	—	Rhodamine B and tetracycline	194.17 mg g^{-1}, and 826.45 mg g^{-1}	—	86
	Mxene	Ti_3AlC_2, HCl, LiF, NH_3/Ar plasma, NH_4HCO_3	—	Other application	—	—	87
Slow evaporation	MOF	1,3,5-Benzenetricarboxylic acid, $Cu(CH_3CO_2)_2·H_2O$	Pyramid	Congo red (CR), methylene blue (MB) and methyl violet (MV)	98% for CR, 96.2% for MB, and 92% for MV	—	88

Diffusion	COF	1,3,5-Triformylphloroglucinol, *n*-hexane, PTSA·H_2O	—	—	—	—	89
		1,3,5-Tri-(4-aminophenyl) benzene (TPB) and 2,5-dimethoxybenzene-1,4-dicarboxaldehyde	Fibrillar and nodular	—	—	—	90
	MOF	Zinc nitrate hexahydrate and 2-methylimidazole	—	Uranium	—	90%	91
Etching	Mxene	Ti_3AlC_2 and HF	Staked sheets	Methylene blue and acid blue 80	—	81% and 62%	92
		Ti_2AlC, LiF, and HCl	2D layered nano sheets	Methylene blue and methyl orange	2460.9 mg g^{-1} and 122.6 mg g^{-1}	—	93
		Ti_2AlC, LiF, and HCl	Layered nano sheets	Pb(II)	Ti_2CTX – 81.5 mg g^{-1}, Ti_2CTX-CS – 93.5 mg g^{-1}, Ti_2CTX-LS – 104.38 mg g^{-1}, Ti_2CTX-EHL – 122.4 mg g^{-1}	—	34
Ball milling	Mxene	3C-SiC	—	Reduction of CO_2 to CH_4	—	90%	94
		Ti_3AlC_2 and HF	Accordion-like structure	Methyl orange	—	90.55%	95

(*continued*)

Table 2.2 (*Continued*)

Method	Material	Linkers	Morphology	Pollutants	Efficiency: Adsorption	Efficiency: Removal	References
Pyrolysis	Biochar	Sewage sludge	Irregular	Bisphenol A	—	80%	96
		Pine cone	—	MB dye and Ni(II)	106.4 mg g^{-1} and 117.7 mg g^{-1}	—	97
		$FeCl_3$ pre-treated biomass (cottonwood)	Rough and porous	As(V)	Reusable adsorbent	Efficient removable	98
		Bamboo	Spherical granular coralline structure	Phenol and methylene blue	169.0 mg g^{-1} and 499.3 mg g^{-1}	—	99
	Biocomposite	Polypropylene, sugar palm starch	Finely textured	—	Other application	—	100
Gasification	Biochar	Biomass gasification effluent, *Rhodococcus opacus*	—	Naphthalene, anthracene, phenanthrene and fluoranthene	—	92.3%, 90.5%, 88.2% and 82.3%	101
	Biocomposite	Coconut shells, $ZnCl_2$	Multiporous structure	AO10	—	100%	102
Torrefaction	Biochar	Corn stover	—	U(VI), methylene blue (MB)	U(VI) adsorption amounts for 250-Air, 250-N2 and 500-N2 were 111.52, 101.57 and 56.21 mg g^{-1} and their MB adsorption amounts were 349.74 mg g^{-1}, 406.82 mg g^{-1}, and 192.67 mg g^{-1}	—	103
		Ammonium persulphate pretreated bamboo	Smooth with pores	Pb(II)	Potential adsorbent	—	104
	Biocomposite	Almond shells, polypropylene–polyethylene	—	—	—	—	105

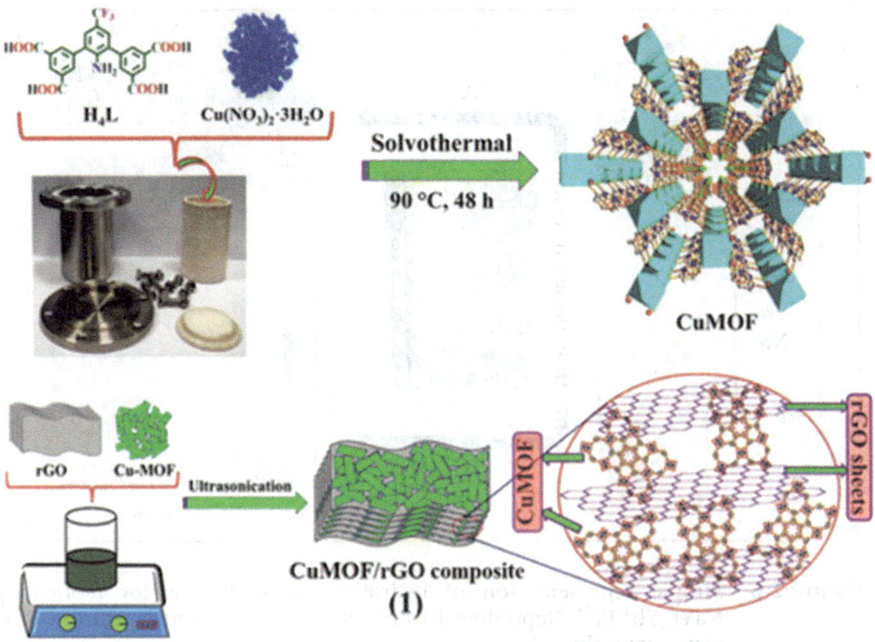

Figure 2.5 Route way of solvothermal synthesis for copper MOF. Reproduced from ref. 108 with permission from American Chemical Society, Copyright 2019.

2.4.2 Hydrothermal

The term "hydrothermal" refers to the processes used to crystallize compounds from highly heated aqueous solutions at high vapour pressures. This technique is preferably used in obtaining a single crystal depending on its solubility in water applied under intense pressure. Hydrothermal is similar to solvothermal which replaces the medium with water. The crystal development process is carried out in a hot, sealed apparatus called an autoclave, made of a steel pressure vessel that receives a water supply and has opposite ends maintained at a specific temperature. The added precursors solute dissolves at the hotter end while it is deposited on a seed crystal at the cooler end, where it grows into the desired crystal.[109] There are autoclaves made of glass with thick walls that can stand temperatures up to 300 °C and 10 bar. Under hydrothermal circumstances, many different types of compounds, including elements, simple and complex oxides, tungstates, molybdates, carbonates, silicates, germanates, *etc.*, have been created. Synthetic crystals with commercial value namely quartz, diamonds, and other single crystals are frequently grown by this process. For instance, emeralds, rubies, quartz, alexandrite, and other crystals have been successfully developed through hydrothermal synthesis. The technique has proven to be incredibly effective in searching for novel compounds with particular physical properties and

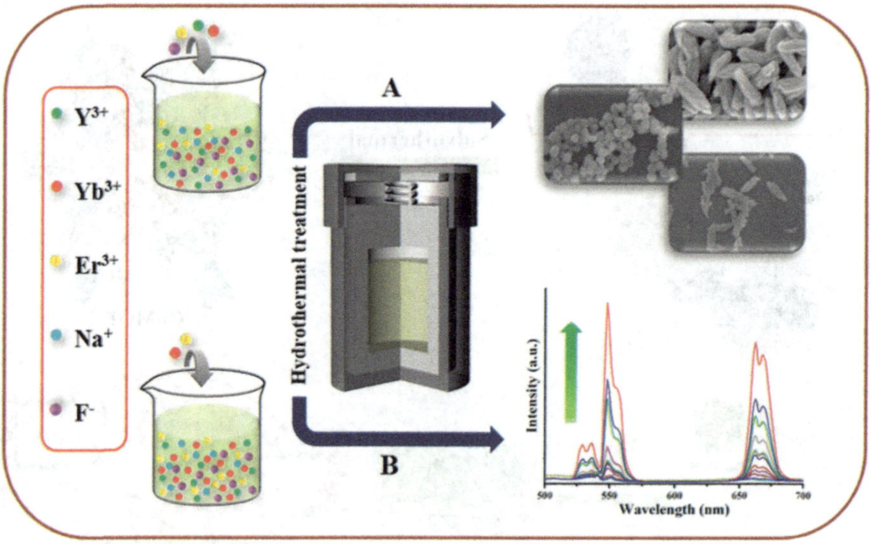

Figure 2.6 Simple representation of hydrothermal treatment for fabricating $NaYF_4Yb^{3+}Er^{3+}$. Reproduced from ref. 111 with permission from Elsevier, Copyright 2019.

in the systematic physicochemical analysis of complex multicomponent systems under high pressures and temperatures.[110] This technique is most commonly used in MOF, biochar, and biocomposite production, which can occur with use of less toxic and non-toxic solvents. These materials are eco-friendly due to the use of water as a medium and minimal loss of product (Figure 2.6).

2.4.3 Microwave-assisted Method

Microwave-assisted synthesis is typically carried out in conjunction with a tried-and-true synthesis technique. This method is based on the idea that material dipoles can be aligned in an external field by being excited by microwave electromagnetic radiations. This approach is very beneficial as the synthesis process can be adjusted to obtain the yield. By using a microwave-assisted synthesis technique, there can be reacquired size and shape.[112] Internal heat may be produced during the process of synchronizing or orienting molecules by the external electrical field, which decreases the processing time and energy requirement. The heating uniformity of microwaves can be significantly shortened.

2.4.4 Sonochemical

In sonochemical synthesis powerful ultrasonic radiation at about 20 kHz–10 MHz is used to make molecules undergo a chemical change. Hot spots produced by sonochemistry can reach extremely high temperatures (5000–25 000 K), pressures as high as 1000 atm, and heating and cooling rates greater than 1011 K s^{-1}.[11] When high-intensity ultrasound is employed, it causes chemical and physical changes that can be used to create or alter a variety of nanostructured materials. Acoustic cavitation is the underlying concept that drives the alteration of nanostructures in the sonochemical process.[113] There are three types of sonochemical reactions:

(i) homogeneous sonochemistry of liquids,
(ii) heterogeneous sonochemistry of liquid–liquid or solid–liquid systems, and
(iii) sonocatalysis using ultrasound to catalyse or speed up a chemical reaction.

The same cavitation mechanisms that cause homogenous sonochemistry also lead to sonoluminescence. In mixed-phase synthesis, materials chemistry, and biological applications, the chemical amplification of processes by ultrasound has been studied. Chemical reactions are not observed when solids or solid–gas systems are exposed to ultrasonic radiation because cavitation can only happen in liquids. Some water contaminants, particularly chlorinated organic molecules, are susceptible to sonochemical destruction.[114]

2.4.5 Techniques for Biochar Preparation

2.4.5.1 *Pyrolysis*

Heating organic materials such as biomass in the absence of oxygen is termed pyrolysis. Generally, biomass is pyrolyzed at temperatures beyond 500 °C, which is hot enough to break down biopolymer bonds. Since combustion requires oxygen, pyrolysis does not produce burned products of biomass. When biomass is pyrolyzed it produces four products: biochar, bio-oil, a liquid, and syngas. The compositions of feedstock and processing parameters are the two factors which influence the amount of these products. The production of bio-oil is at its maximum under ideal pyrolysis conditions such as a temperature close to 500 °C and a high heating rate (1000 °C s^{-1}). Under these circumstances, a typical biomass feedstock can provide 60–70 weight percentage yields of bio-oil and 15–25 weight percentage yields of biochar, syngas making up the remaining 10–15 weight percentage. Biochar is typically the primary by-product of slow pyrolysis methodologies, which use slower heating rates. Because of the production of sufficient energy needed to power the pyrolysis process by the burning of some biochar or oil the pyrolysis techniques are termed self-sustaining[115] (Figure 2.7).

2.4.5.2 *Gasification*

The thermochemical conversion of thermal energy in carbonaceous materials into syngas (a mixture of CO and H_2), tars, and biochar at ambient temperatures (over 500 °C) and in the absence of oxygen is referred

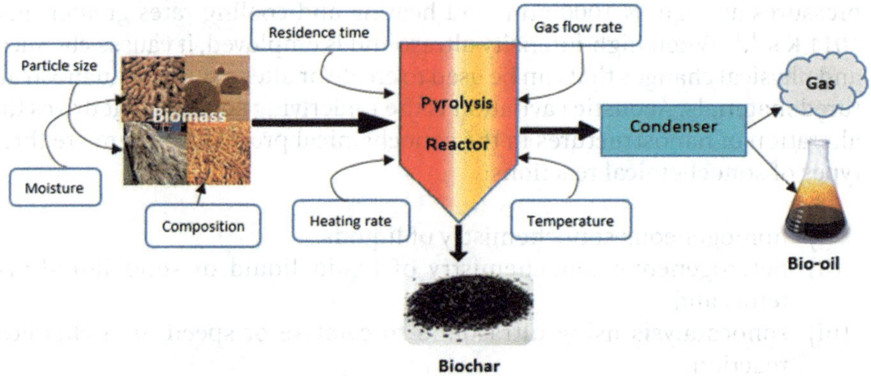

Figure 2.7 Ideology of extracting biochar from biomass. Reproduced from ref. 116 with permission from Springer Nature, Copyright 2020.

to as gasification. Drying, pyrolysis, partial oxidation, and reduction are the typical four phases of gasification. In comparison with other thermochemical processes such as pyrolysis, the gasification technique yields a lower amount of biochar. This is because of the partial oxidation conditions of carbon that induce it to be converted into carbon monoxide (CO). A variety of gasification operational conditions are adopted to effectively maximize the energy (*i.e.*, the production of syngas) from multiple carbonaceous feedstocks. The gasification process deliberately hinders the formation of biochar as a by-product to optimize energy recovery (*i.e.*, high yield of syngas). As a result, this inherently restricts the gasification's operational parameters. Greater energy production per mass of carbonaceous material, decreased power consumption, enhanced heat generation, and improved chemical and textural features of biochar due to partial oxidation during gasification are all benefits of creating biochar by gasification.[117]

2.5 Top-down and Bottom-up Approaches

There are numerous ways to create nanomaterials. These techniques are divided into two groups: top-down and bottom-up approaches. The techniques are categorized depending on the stage of the raw material. The initial substance is solid in the top-down class of procedures, whereas the starting material in bottom-up techniques is either gaseous or liquid. Top-down approaches are subtractive processes that break down large beginning materials into smaller, nanoscale ones. Bottom-up approaches are additive processes that begin with precursor molecules or atoms and then combine to create nanoscale structures. Nanostructures are constructed using a bottom-up strategy, atom-by-atom or molecule-by-molecule formulation in these kinds of approaches.

2.5.1 Ball-milling

Ball milling, commonly referred to as mechanical grinding, is an easy and well-liked technique. Here, the components are ground into ultra-fine powders. One of the most important industrial techniques for creating nanomaterials is ball milling. The kinetic energy produced by the moving balls is used to break down the chemical bonds of the larger molecules and reduce the particle size to nanoscale.[118] This approach has many benefits, including a scalable architecture, low maintenance requirements, and a compact design. It operates on the idea of impact. The balls impact as they fall from the top of the chamber holding the source material resulting in the size decrease.

2.5.2 Etching

Chemical etching is the process of removing a surface layer from a metal or plastic surface through chemical erosion. One or more chemical reactions take place throughout this phase, consuming the original reactants and creating new species (Table 2.2).

2.6 Challenges and Future Prospects

Although the mentioned materials are efficient in the removal of pollutants with functionalization, they are associated with a number of challenges. Firstly, the MXene has obstacles due to high ionic diffusion resistance that reduces the capability of the material in respective applications. Also, the stacking of MXene single layer limits the cation's intercalation which reduces the efficiency of the material. On functionalization with carboxyl, amine, acid, and polymer coating on MXene can have property enhancement such as increased surface area that results in increased adsorption of contaminant molecules.[119] MOF faces a few challenges such as stability, fatigue, recyclability (<10 cycles), and reusability. Most MOFs are produced from toxic metals which can also be a drawback. Some major factors for the utilization of MOF in water pollutant removal are (i) composition of contaminants, (ii) real concentration range of organic and various pollutants, (iii) impact of contaminant species targeted for removal efficiency, and (iv) pH and temperature of wastewater.[120] The usage of the covalent framework in the removal of pollutants is not as stable as the use in energy and sensor fields. The drawbacks associated with COF are the toxic organic linkers used during the primary stage of synthesizing. The stability of COF is also a neglectable factor. But from a future perspective, the functionalization of COF can result in efficient property changes based on stability and toxicity. The concept of using biochar and biocomposite for the removal of pollutants is an excellent alternative for replacing modern and toxic advanced synthetic materials. These adsorption materials also have drawbacks such as the weak force of attraction between the composite molecules and thus, the composite may break and disperse in the contaminant water which may further lead to filtration techniques for the reuse of wastewater. The advantage of producing adsorbent material based on biochar and biocomposite can reduce the accumulation of biowaste on the land. This biowaste has a higher porous nature and pore volume which also aids in faster adsorption rate. The rising scarcity of water scarcity can be overcome by the treatment of advanced materials due to the advantages of high surface area, porosity, and thermal stability. Surface modification on emerging advanced materials as discussed above can overcome a few of the mentioned drawbacks. Also, following modern techniques for synthesizing advanced material can produce unexpected yields with reduced manpower. Thus, modern advanced materials and bioadsorbents are excellent materials for the adsorption of pollutants from water.

2.7 Conclusion

Pollutants are emerging in the environment at a higher rate and their toxicities are high. The enduring nature results in bioaccumulation, which can eventually constitute ecosystem risk. Thus, the necessity of removing pollutants from water sources requires advanced materials with high-end efficiency. This chapter considered types of advanced materials for pollutant removal such as metal–organic framework, MXene, covalent organic framework, and biochar along with their synthesis and functionalization methods. The vital technologies for synthesizing the advanced materials and their efficiency in pollutant removal are also discussed. The chapter also focuses on the functionalization effect on materials to improve their properties such that it can lead to high removal efficiency. In conclusion, the comprehensive insight into the methodology, surface modification *via* suitable functionalization techniques and various methods of synthesizing the advanced material will aid in optimizing the accurate material for performing the adsorption of pollutants from water sources depending on their properties.

References

1. M. Haseena, M. Faheem Malik, A. Javed, S. Arshad, N. Asif, S. Zulfiqar and J. Hanif, *Env. Risk Assess. Rem.*, 2017, **1**(3), 16–19.
2. I. C. Vasilachi, D. M. Asiminicesei, D. I. Fertu and M. Gavrilescu, *Water*, 2021, 13.
3. R. Saxena, M. Saxena and A. Lochab, *ChemistrySelect*, 2020, **5**, 335–353.
4. K. Adesina Adegoke, O. Samuel Agboola, J. Ogunmodede, A. Oluyomi Araoye and O. Solomon Bello, *Mater. Chem. Phys.*, 2020, **253**, 123246.
5. M. Khatami and S. Iravani, *Comments Inorg. Chem.*, 2021, **41**, 213–248.
6. P. Kumar, V. Bansal, K. H. Kim and E. E. Kwon, *J. Ind. Eng. Chem.*, 2018, **62**, 130–145.
7. I. Ihsanullah, *Chem. Eng. J.*, 2020, 388.
8. D. Jaspal and A. Malviya, *Chemosphere*, 2020, 246.
9. F. Gao, *ChemistrySelect*, 2019, **4**, 6805–6811.

10. T. Chen, B. Li, W. Huang, C. Lin, G. Li, H. Ren, Y. Wu, S. Chen, W. Zhang and H. Ma, *Sep. Purif. Technol.*, 2021, **256**, 117787.
11. A. R. Bagheri, N. Aramesh, F. Sher and M. Bilal, *Chemosphere*, 2021, **270**, 129523.
12. X. Guan, F. Chen, Q. Fang and S. Qiu, *Chem. Soc. Rev.*, 2020, **49**, 1357–1384.
13. N. Huang, P. Wang and D. Jiang, *Nat. Rev. Mater.*, 2016, 1.
14. W.-X. Wu, F. Li, B.-J. Yao, L.-G. Ding, J.-L. Kan, F. Liu, G.-Y. Zhao, S. Wang and Y.-B. Dong, *J. Hazard. Mater.*, 2022, **433**, 128831.
15. W. Ma, Q. Zheng, Y. He, G. Li, W. Guo, Z. Lin and L. Zhang, *J. Am. Chem. Soc.*, 2019, **141**, 18271–18277.
16. E. M. Dias and C. Petit, *J. Mater. Chem. A*, 2015, **3**, 22484–22506.
17. Q. Gao, J. Xu and X. H. Bu, *Coord. Chem. Rev.*, 2019, **378**, 17–31.
18. K. Wang, H. Huang, W. Xue, D. Liu, X. Zhao, Y. Xiao, Z. Li, Q. Yang, L. Wang and C. Zhong, *Chem. Commun.*, 2012, **48**, 8437–8439.
19. R. M. Rego, G. Kuriya, M. D. Kurkuri and M. Kigga, *J. Hazard. Mater.*, 2021, 403.
20. S. Wang, Y. Lv, Y. Yao, H. Yu and G. Lu, *Inorg. Chem. Commun.*, 2018, **93**, 56–60.
21. J. A. Kumar, P. Prakash, T. Krithiga, D. J. Amarnath, J. Premkumar, N. Rajamohan, Y. Vasseghian, P. Saravanan and M. Rajasimman, *Chemosphere*, 2022, **286**, 131607.
22. L. Damptey, B. N. Jaato, C. S. Ribeiro, S. Varagnolo, N. P. Power, V. Selvaraj, D. Dodoo-Arhin, R. V. Kumar, S. P. Sreenilayam, D. Brabazon, V. Kumar Thakur and S. Krishnamurthy, *Glob. Chall.*, 2022, **6**, 2100120.
23. J. U. Ani, K. G. Akpomie, U. C. Okoro, L. E. Aneke, O. D. Onukwuli and O. T. Ujam, *Appl. Water Sci.*, 2020, 10.
24. M. S. Reza, C. S. Yun, S. Afroze, N. Radenahmad, M. S. A. Bakar, R. Saidur, J. Taweekun, A. K. Azad and J. Arab, *Basic Appl. Sci.*, 2020, **27**, 208–238.
25. S. Thakur, in *Materials Today: Proceedings*, Elsevier Ltd, 2020, vol. 37, pp. 3305–3309.
26. M. Ahmad, A. U. Rajapaksha, J. E. Lim, M. Zhang, N. Bolan, D. Mohan, M. Vithanage, S. S. Lee and Y. S. Ok, *Chemosphere*, 2014, **99**, 19–33.
27. W. Jiang, W. R. Cui, R. P. Liang and J. D. Qiu, *J. Hazard. Mater.*, 2021, **413**, 125302.
28. A. Ghahari, H. Raissi and F. Farzad, *J. Taiwan Inst. Chem. Eng.*, 2021, **125**, 15–22.
29. J. Li, X. Yang, C. Bai, Y. Tian, B. Li, S. Zhang, X. Yang, S. Ding, C. Xia, X. Tan, L. Ma and S. Li, *J. Colloid Interface Sci.*, 2015, **437**, 211–218.
30. H. Hoseinzadeh, B. Hayati, F. Shahmoradi Ghaheh, K. Seifpanahi-Shabani and N. M. Mahmoodi, *Mater. Res. Bull.*, 2021, **142**, 111408.
31. E. Wu, Y. Li, Q. Huang, Z. Yang, A. Wei and Q. Hu, *Chemosphere*, 2019, **233**, 327–335.
32. H. S. Far, M. Hasanzadeh, M. S. Nashtaei, M. Rabbani, A. Haji and B. Hadavi Moghadam, *ACS Appl. Mater. Interfaces*, 2020, **12**, 25294–25303.

33. D. Gan, Q. Huang, J. Dou, H. Huang, J. Chen, M. Liu, Y. Wen, Z. Yang, X. Zhang and Y. Wei, *Appl. Surf. Sci.*, 2019, 144603.
34. S. Wang, Y. Liu, Q. F. Lü and H. Zhuang, *J. Mol. Liq.*, 2020, **297**, 111810.
35. X. Zeng, Y. Wang, X. He, C. Liu, X. Wang and X. Wang, *J. Environ. Chem. Eng.*, 2021, **9**, 106203.
36. J. R. J. Zaeni, J. W. Lim, Z. Wang, D. Ding, Y. S. Chua, S. L. Ng and W. Da Oh, *Sep. Purif. Technol.*, 2020, **241**, 116702.
37. J. Choi, W. Won and S. C. Capareda, *Ind. Crops Prod.*, 2019, **137**, 672–680.
38. J. Deng, X. Li, X. Wei, Y. Liu, J. Liang, Y. Shao, W. Huang and X. Cheng, *Bioresour. Technol.*, 2020, **310**, 123438.
39. I. F. Nata, D. R. Wicakso, A. Mirwan, C. Irawan, D. Ramadhani, and Ursulla, *J. Environ. Chem. Eng.*, 2020, **8**, 104339.
40. R. Nowruzi, M. Heydari and V. Javanbakht, *Int. J. Biol. Macromol.*, 2020, **147**, 209–216.
41. J. Hao, P. Chen, Q. Zhang, P. Chen, X. Zheng, Y. Wu, D. Ma, D. Wei, H. Liu, G. Liu and W. Lv, *Environ. Sci.: Nano*, 2019, **6**, 3374–3387.
42. L. Huang, R. Shen, R. Liu and Q. Shuai, *J. Hazard. Mater.*, 2020, **392**, 122320.
43. A. Bagheri, H. Hoseinzadeh, B. Hayati, N. M. Mahmoodi and E. Mehraeen, *J. Environ. Chem. Eng.*, 2020, 104590.
44. N. M. Mahmoodi, M. Oveisi, A. Taghizadeh and M. Taghizadeh, *J. Hazard. Mater.*, 2019, **368**, 746–759.
45. G. Yang, X. Hu, J. Liang, Q. Huang, J. Dou, J. Tian, F. Deng, M. Liu, X. Zhang and Y. Wei, *J. Hazard. Mater.*, 2021, **419**, 126220.
46. P. Zhang, L. Wang, K. Du, S. Wang, Z. Huang, L. Yuan, Z. Li, H. Wang, L. Zheng, Z. Chai and W. Shi, *J. Hazard. Mater.*, 2020, **396**, 122731.
47. J. Buates and T. Imai, *J. Water Process Eng.*, 2020, **37**, 101508.
48. A. Fernandez-Sanroman, V. Acevedo-García, M. Pazos, M. A. Sanromán and E. Rosales, *J. Cleaner Prod.*, 2020, **271**, 122436.
49. J. Ge, J. Xiao, L. Liu, L. Qiu and X. Jiang, *J. Porous Mater.*, 2016, **23**, 791–800.
50. W. Ji, Y. S. Guo, H. M. Xie, X. Wang, X. Jiang and D. S. Guo, *J. Hazard. Mater.*, 2020, **397**, 122793.
51. F. Zou, R. Yu, R. Li and W. Li, *ChemPhysChem*, 2013, **14**, 2825–2832.
52. M. E. Mahmoud, M. F. Amira, S. M. Seleim and A. K. Mohamed, *J. Hazard. Mater.*, 2020, **381**, 120979.
53. J. He, Y. Zhang, X. Zhang and Y. Huang, *Sci. Rep.*, 2018, **8**, 5159.
54. Y. Li, Y. Liu, W. Gao, L. Zhang, W. Liu, J. Lu, Z. Wang and Y. J. Deng, *CrystEngComm*, 2014, **16**, 7037–7042.
55. J. Ge, Z. Wu, X. Huang and M. Ding, *J. Nanomater.*, 2019, 4035075.
56. Q. Wu, Y. Ren, P. Li, Y. Wang, Z. Yang, K. Qu, G. Qi, S. Chen and F. Wu, *Appl. Phys. A: Mater. Sci. Process.*, 2021, **127**, 360.
57. J. Qu, M. Dong, S. Wei, Q. Meng, L. Hu, Q. Hu, L. Wang, W. Han and Y. Zhang, *Carbohydr. Polym.*, 2020, **250**, 117003.
58. S. Anfar, M. Zbair, H. A. Ahsiane, A. Jada and N. E. Alem, *RSC Adv.*, 2020, **10**, 11371–11380.

59. A. Nayak, B. Bhushan, V. Gupta and S. Kotnala, *J. Ind. Eng. Chem.*, 2024, **288**, 119796.
60. J. Qu, Q. Meng, X. Lin, W. Han, Q. Jiang, L. Wang, Q. Hu, L. Zhang and Y. Zhang, *Sci. Total Environ.*, 2021, **752**, 141854.
61. F. Tamaddon, M. H. Mosslemin, A. Asadipour, M. A. Gharaghani and A. Nasiri, *Int. J. Biol. Macromol.*, 2020, **154**, 1036–1049.
62. A. Mellah, S. P. S. Fernandes, R. Rodríguez, J. Otero, J. Paz, J. Cruces, D. D. Medina, H. Djamila, B. Espiña and L. M. Salonen, *Chem. - Eur. J.*, 2018, **24**, 10601–10605.
63. H. Tian, J. Peng, T. Lv, C. Sun and H. He, *J. Solid State Chem.*, 2018, **257**, 40–48.
64. R. Guo, X. Cai, H. Liu, Z. Yang, Y. Meng, F. Chen, Y. Li and B. Wang, *Environ. Sci. Technol.*, 2019, **53**, 2705–2712.
65. H. Li, X. Cao, C. Zhang, Q. Yu, Z. Zhao, X. Niu, X. Sun, Y. Liu, L. Ma and Z. Li, *RSC Adv.*, 2017, **7**, 16273–16281.
66. S. Luo, R. Wang, J. Yin, T. Jiao, K. Chen, G. Zou, L. Zhang, J. Zhou, L. Zhang and Q. Peng, *ACS Omega*, 2019, **4**, 3946–3953.
67. M. Afshari and M. Dinari, *J. Hazard. Mater.*, 2020, **385**, 121514.
68. M. Afshari, M. Dinari, K. Zargoosh and H. Moradi, *Ind. Eng. Chem. Res.*, 2020, **59**, 9116–9126.
69. Y. Sun, M. Chen, H. Liu, Y. Zhu, D. Wang and M. Yan, *Appl. Surf. Sci.*, 2020, **525**, 146614.
70. N. M. Mahmoodi, M. Taghizadeh and A. Taghizadeh, *J. Mol. Liq.*, 2019, **277**, 310–322.
71. Y. Li, J. Wang, Z. Huang, C. Qian, Y. Tian and Y. Duan, *J. Environ. Chem. Eng.*, 2021, **9**, 106012.
72. C. Cui, R. Guo, H. Xiao, E. Ren, Q. Song, C. Xiang, X. Lai, J. Lan and S. Jiang, *Appl. Surf. Sci.*, 2020, **505**, 144595.
73. A. Shahzad, K. Rasool, M. Nawaz, W. Miran, J. Jang, M. Moztahida, K. A. Mahmoud and D. S. Lee, *Chem. Eng. J.*, 2018, **349**, 748–755.
74. S. Kandambeth, V. Venkatesh, D. B. Shinde, S. Kumari, A. Halder, S. Verma and R. Banerjee, *Nat. Commun.*, 2015, **6**, 6786.
75. N. Hammi, S. El Hankari, N. Katir, N. Marcotte, K. Draoui, S. Royer and A. El Kadib, *Microporous Mesoporous Mater.*, 2020, **306**, 110429.
76. S. Salehi, S. Mandegarzad and M. Anbia, *J. Alloys Compd.*, 2020, **812**, 152051.
77. X. Xiao, H. Yu, H. Jin, M. Wu, Y. Fang, J. Sun, Z. Hu, T. Li, J. Wu, L. Huang, Y. Gogotsi and J. Zhou, *ACS Nano*, 2017, **11**, 2180–2186.
78. L. Huang, N. Mao, Q. Yan, D. Zhang and Q. Shuai, *ACS Appl. Nano Mater.*, 2020, **3**, 319–326.
79. L. Wang, H. Song, L. Yuan, Z. Li, Y. Zhang, J. K. Gibson, L. Zheng, Z. Chai and W. Shi, *Environ. Sci. Technol.*, 2018, **52**, 10748–10756.
80. C. Sun, L. Karuppasamy, L. Gurusamy, H. J. Yang, C. H. Liu, J. Dong and J. J. Wu, *Sep. Purif. Technol.*, 2021, **271**, 118873.
81. R. Abazari and A. R. Mahjoub, *Ultrason. Sonochem.*, 2018, **42**, 577–584.

82. M. Saidi, A. Benomara, M. Mokhtari and L. Boukli-Hacene, *React. Kinet., Mech. Catal.*, 2020, **131**, 1009–1021.
83. E. Moradi, R. Rahimi, V. Safarifard and S. Azari, *Proceedings*, 2019, **41**, 31.
84. M. Tanhaei, A. R. Mahjoub and V. Safarifard, *Ultrason. Sonochem.*, 2018, **41**, 189–195.
85. A. Shahzad, K. Rasool, W. Miran, M. Nawaz, J. Jang, K. A. Mahmoud and D. S. Lee, *ACS Sustainable Chem. Eng.*, 2017, **5**, 11481–11488.
86. J. Wu, Y. Gao, S. Wei, P. Chen, D. Gu, B. Fu and M. Chen, *J. Solid State Chem.*, 2021, **302**, 122350.
87. X. Chen, X. Zhai, J. Hou, H. Cao, X. Yue, M. Li, L. Chen, Z. Liu, G. Ge and X. Guo, *Chem. Eng. J.*, 2021, **420**, 129832.
88. G. I. Dzhardimalieva, R. K. Baimuratova, E. I. Knerelman, G. I. Davydova, S. E. Kudaibergenov, O. V. Kharissova, V. A. Zhinzhilo and I. E. Uflyand, *Polymers*, 2020, **12**, 1024.
89. R. Wang, X. Shi, Z. Zhang, A. Xiao, S. P. Sun, Z. Cui and Y. Wang, *J. Membr. Sci.*, 2019, **586**, 274–280.
90. P. Manchanda, S. Chisca, L. Upadhyaya, V. E. Musteata, M. Carrington and S. P. Nunes, *J. Mater. Chem. A*, 2019, **7**, 25802–25807.
91. B. Yu, G. Ye, J. Chen and S. Ma, *Environ. Pollut.*, 2019, **253**, 39–48.
92. O. Mashtalir, K. M. Cook, V. N. Mochalin, M. Crowe, M. W. Barsoum and Y. Gogotsi, *J. Mater. Chem. A*, 2014, **2**, 14334–14338.
93. B. Sun, X. Dong, H. Li, Y. Shang, Y. Zhang, F. Hu, S. Gu, Y. Wu, T. Gao and G. Zhou, *Sep. Purif. Technol.*, 2021, **272**, 118964.
94. H. Li and J. Sun, *ACS Appl. Mater. Interfaces*, 2021, **13**, 5073–5078.
95. Z. Cao, J. Su, Y. Li, J. Li, Z. Wang, M. Li, B. Fan, G. Shao, H. Wang, H. Xu, R. Zhang and H. Lu, *J. Alloys Compd.*, 2021, **889**, 161771.
96. B. C. Huang, J. Jiang, G. X. Huang and H. Q. Yu, *J. Mater. Chem. A*, 2018, **6**, 8978–8985.
97. S. Dawood, T. K. Sen and C. Phan, *Bioresour. Technol.*, 2017, **246**, 76–81.
98. M. Zhang, B. Gao, S. Varnoosfaderani, A. Hebard, Y. Yao and M. Inyang, *Bioresour. Technol.*, 2013, **130**, 457–462.
99. Z. Li, B. Xing, Y. Ding, Y. Li and S. Wang, *Chin. J. Chem. Eng.*, 2020, **28**, 2872–2880.
100. M. M. Harussani, S. M. Sapuan, U. Rashid and A. Khalina, *Polymers*, 2021, **13**(11), 1707.
101. L. Goswami, N. A. Manikandan, J. C. R. Taube, K. Pakshirajan and G. Pugazhenthi, *Environ. Sci. Pollut. Res.*, 2019, **26**, 25154–25166.
102. Y. Zheng, D. Chen, N. Li, Q. Xu, H. Li, J. He and J. Lu, *Chemosphere*, 2017, **179**, 139–147.
103. L. Li, M. Yang, Q. Lu, W. Zhu, H. Ma and L. Dai, *Bioresour. Technol.*, 2019, **294**, 122142.
104. Y. Shen, J. Z. Guo, L. Q. Bai, X. Q. Chen and B. Li, *Bioresour. Technol.*, 2021, **323**, 124616.
105. Z. McCaffrey, L. Torres, S. Flynn, T. Cao, B. S. Chiou, A. Klamczynski, G. Glenn and W. Orts, *Ind. Crops Prod.*, 2018, **125**, 425–432.

106. T. Ingsel and R. K. Gupta, in *Metal-Organic Framework-Based Nanomaterials for Energy Conversion and Storage Micro and Nano Technologies*, 2022, pp. 11–33.
107. S. Choudhury, S. Paul, S. Goswami and K. Deb, in *Advances in Nanotechnology-Based Drug Delivery Systems Nanotechnology in Biomedicine*, 2022, pp. 21–44.
108. A. K. Gupta, M. Saraf, P. K. Bharadwaj and S. M. Mobin, *Inorg. Chem.*, 2019, **58**, 9844–9854.
109. E. Mariani, C. Sabelli and L. Vergelli, *Operating Experience with Peak Generation Plants in a Hydro-Thermal Electric Power System*, 1979.
110. E. B. Denkbaş, E. Çelik, E. Erdal, D. Kavaz, Ö. Akbal, G. Kara and C. Bayram, in *Nanobiomaterials in Drug Delivery: Applications of Nanobiomaterials*, Elsevier Inc., 2016, pp. 285–331.
111. R. Rafique, S. H. Baek, L. M. T. Phan, S. J. Chang, A. R. Gul and T. J. Park, *Mater. Sci. Eng., C*, 2019, **99**, 1067–1074.
112. M. S. Dahiya, V. K. Tomer and S. Duhan, in *Applications of Nanocomposite Materials in Drug Delivery*, Elsevier, 2018, pp. 737–760.
113. J. González-García, V. Sáez, I. Tudela, M. I. Díez-Garcia, M. D. Esclapez and O. Louisnard, *Water*, 2010, **2**, 28–74.
114. K. S. Suslick, T. Hyeon, M. Fang and A. A. Cichowlas, *Sonochemical Synthesis of Nanostructured Catalysts*, 1995.
115. H. Zhou, Y. Long, A. Meng, Q. Li and Y. Zhang, *Thermochim. Acta*, 2013, **566**, 36–43.
116. A. K. Sakhiya, A. Anand and P. Kaushal, *Biochar*, 2020, **2**, 253–285.
117. S. You, Y. S. Ok, S. S. Chen, D. C. W. Tsang, E. E. Kwon, J. Lee and C. H. Wang, *Bioresour. Technol.*, 2017, **246**, 242–253.
118. M. Y. S. O. Kumar, X. Xiong, Z. Wan, Y. Sun, D. C. W. Tsang, J. Gupta, B. Gao, X. Cao, J. Tang and Y. S. Ok, *Bioresour. Technol.*, 2020, **312**, 123613.
119. S. Nahirniak, A. Ray and B. Saruhan, *Batteries*, 2023, **9**, 126.
120. S. Rojas and P. Horcajada, *Chem. Rev.*, 2020, **120**, 8378–8415.

CHAPTER 3

Membranes for the Removal of Endocrine Disrupting Compounds from Aqueous Environments

KHAIRUL ANWAR MOHAMAD SAID[a,b,c,*], CLARA SARI[a] AND MD REZAUR RAHMAN[a,c]

[a]Department of Chemical Engineering and Energy Sustainability, Faculty of Engineering, Universiti Malaysia Sarawak (UNIMAS), Malaysia; [b]Advanced Membrane and Material Research Unit (AMMRU), Faculty of Engineering, Universiti Malaysia Sarawak (UNIMAS), Malaysia; [c]UNIMAS Water Centre, Faculty of Engineering, Universiti Malaysia Sarawak (UNIMAS), Malaysia
E-mail: mskanwar@unimas.my

3.1 Introduction

Endocrine-disrupting chemicals or compounds (EDC) are varied arrays of compounds that can disrupt the typical operation of the endocrine system, which controls the production and regulation of hormones in the body.[1] These compounds have the ability to imitate, obstruct, or disrupt the function of natural hormones, resulting in various negative impacts on human health and the environment.[2] EDCs encompass a range of chemical categories, including pharmaceuticals, pesticides, substances utilized in the plastics sector and consumer goods, industrial waste items, contaminants, and

Chemistry in the Environment Series No. 14
Advanced Materials for Emerging Water Pollutant Removal
Edited by Pei Sean Goh, Devagi Kanakaraju, Anwar Iqbal and Ahmad Fauzi Ismail

Published by the Royal Society of Chemistry, www.rsc.org

certain naturally occurring plant components.[3] EDCs encompass both natural molecules like phytoestrogens and manufactured chemicals such as pesticides, plastics, and medications.

The common technique employed to eliminate chemical pollutants from drinking water treatment systems is not highly efficient. Research has shown that flocculation or coagulation processes are able to remove less than 40% of EDC.[4] UV irradiation, ozonation, and membrane treatment are advanced methods that can enhance water purification by eliminating EDCs.[5] Compared to other advanced technologies that may discharge treatment by-products into the water, membranes are highly effective at reducing the concentration of EDCs.[6] Membrane techniques such as microfiltration (MF), ultrafiltration (UF), nanofiltration (NF), and reverse osmosis (RO) are commonly used in water treatment facilities.[7]

3.2 Toxicity of EDC

3.2.1 Source and Groups of EDCs

EDCs, according to the description by the US Environmental Protection Agency can be categorized as a substance that is not produced by an organism with the ability to manipulate the hormone metabolism.[8] The mechanism by which EDCs disrupt hormones starts by them mimicking the hormone structure that allows them to bind the receptor. The binding of EDCs to the receptor blocks the actual hormone thus leading to instability especially to the organisms endocrine, immune, and nervous systems.[9] The disruption in the vital function could cause several diseases mainly cancer, diabetes, and infertility.[10] As well as diseases, EDCs may contribute to genetic changes by modifying the DNA through acetylation and methylation which could still occur even at low concentrations.[11] A verbatim description by the World Health Organization (WHO) regarding endocrine disrupting compounds is: "group of exogenous compounds that interfere with endocrine system functioning and has adverse health effects on organ-ism and its subpopulation/progeny".[12]

Another name for EDCs is endocrine modulator and environmental hormone which can be categorized as either natural or synthetic depending on the synthesis pathways. Some of the synthetic endocrine modulators are bisphenol A, polychlorinated biphenyls, and triclosan while the natural source is phytoestrogen which can be found in soy milk. Meanwhile, the EDCs that are responsible for interfering with estrogen hormone are xenoestrogen which can be sourced from bisphenol A and soy milk. Both EDCs are known to mimic the estrogen hormone which allow it to bind to the hormone receptor that control the reproductive system. The binding of EDCs to the receptor in the case of xenoestrogen will induce irregular development, DNA modification, and the alteration of the placenta.[13] In the US, based on population sampling from urine analysis, it is approximated that 92% of the US population has BPA in their system. This is worrying because even at low concentration, EDC could alter the DNA therefore increasing the public health risk.[14] In the last decade, much research has been conducted focusing on the interaction of EDCs with organisms at molecular level and the associated hazard.[15] This sudden increase in attention is due to widespread presence of EDCs in our day-to-day life from cosmetics to food containers.[16]

The source of EDCs can be traced as a result of advancement in technology, revolution in manufacturing, and urbanization. The primary sources of EDCs are effluent from hospitals, sewage plants, and livestock industries despite them being treated before discharge. The effluent discharge flows into water resources such as rivers and groundwater. Another concern was the discharged of untreated effluent rich with EDCs into the soil and water resource which contribute to the build-up of EDCs in the environment that eventually affect marine life and water supply.

Each EDC has a different chemical structure but it is noteworthy that every single EDC has a similar effect. At this time, the water treatment technologies designed for treating EDCs must be designed specifically for each type of EDC and no water treatment method can treat all EDCs. In an increasing number of xenobiotic types of chemical produced, its release into the environment has made it hard to estimate the environmental impact of EDCs. Recent events have made it an urgent issue for academia and industry to research and propose a procedure or method that could curb the spread of EDCs in the environment. EDCs can be categorized into several classes based on their chemical structures. For instance, pharmaceuticals, pesticide products, and personal care goods can be classified into three groups that contribute to endocrine disruption.

3.3 Conventional EDC Treatment Method

Endocrine disrupting compounds along with their by-products are the main contaminants in effluent from hospitals, sewage plants, and wastewater from livestock industries. The treatment process for EDC rich wastewater is categorized into three main types: (1) biological, (2) chemical, and (3) physical treatments. These three treatments can be combined as shown in Figure 3.1 for the better removal of particles and production of better quality wastewater.

Common wastewater treatments are split into preliminary, primary secondary, and tertiary treatments in treating wastewater effluents.[17] Preliminary treatment involves equalization and mixing all wastewater streams from different stages in the manufacturing process. This step ensures that the collected wastewater has uniform characteristics in terms of pH, temperature, and pollution load. Primary treatment involves screening, sedimentation, and floatation of effluents; however, some suspended fine elements are not removed, which will be further processed in secondary treatment. Secondary treatment is carried out with help of fungi, bacteria, and other microbes to stabilize the color and dissolved organic material in the wastewater. Tertiary

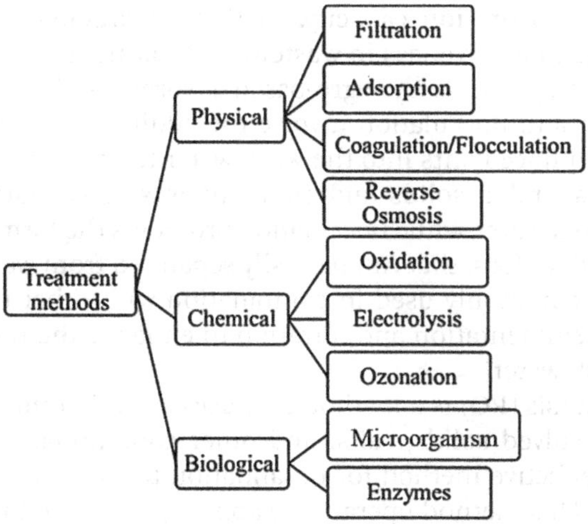

Figure 3.1 Treatment methods for EDC removal in wastewater.

treatment is generally used to remove organic compounds by adsorption and dissolved solids by membrane filtration techniques.

3.3.1 Physical Treatment

Physical wastewater treatment processes are essential for the removal of solid particles, contaminants, and pollutants from wastewater. These processes, such as membrane filtration, adsorption, and coagulation, play a crucial role in achieving efficient and effective wastewater treatment. By combining physical methods with chemical and biological processes, conventional wastewater treatment methods can meet the required effluent discharge limits and ensure the protection of the environment and human health.

Filtration in physical wastewater treatment involves the separation of solid particles and contaminants from wastewater by passing it through a medium with a certain level of porosity. There are various filtration techniques, such as membrane filtration, microfiltration, and nanofiltration.[18] Membrane filtration including reverse osmosis is widely used as it utilizes a semi-permeable membrane to separate suspended solids, dissolved solids, and other contaminants found in the wastewater. Membrane filtration is effective in removing particles, microorganisms, and dissolved particles from wastewater which produces high-quality effluent.

Adsorption involves the attachment of contaminants to a solid surface, such as activated carbon or other types of adsorbents, to remove them from the wastewater. This method is effective in removing various pollutants, such as dyes, heavy metals, and other organic compounds from wastewater. The most commonly used adsorbent is activated carbon due to its high surface area and high adsorption capacity.[19] Other adsorbents which have been studied for their effectiveness in wastewater treatment are graphene-oxide-chitosan hybrid hydrogels and agro-based adsorbents.[20]

Coagulation and flocculation involve the addition of chemicals, such as coagulants and flocculants into the wastewater to promote aggregation and settling of suspended solids and contaminants. Coagulation destabilizes particles in wastewater, while flocculation promotes the formation of clumps of particles called flocs, that can be easily separated from water.[21] Both these processes are commonly used in combination with other treatment methods, such as sedimentation and filtration to enhance the removal of pollutants from wastewater.

Reverse osmosis (RO) is a method that uses a semi-permeable membrane to remove dissolved solids, salts, and other contaminants from wastewater. RO is an effective method for desalination purpose as it produces high quality water. This method operates by applying pressure to the wastewater, which forces the water molecules to pass through the membrane while leaving the dissolved solids and contaminants behind.[18] Reverse osmosis is commonly combined with other treatment techniques, such as coagulation and flocculation, enhancing the removal of particles from wastewater.

However, these methods have their own limitations. Filtration has limited effectiveness in removing dissolved contaminants and small particles such as organic compounds with low molecular weight. Filtration that uses a membrane is prone to membrane fouling which will reduce the efficiency of filtration and will require frequent cleaning or replacements of membranes. Next, adsorption capacity is affected by the presence of other substances in the wastewater, such as dissolved organic matter, which can compete for adsorption sites. This is due to its limited selectivity for specific contaminants, leading to the potential of removing both the targeted and non-targeted compounds.[22] Furthermore, the adsorbents used can be saturated and will need to be regenerated or replaced.

Coagulation and flocculation are highly affected by the composition and pH of the wastewater. These methods are limited to certain contaminants, such as dissolved organic matter. The main issue in using these methods will be their disposal, where they generate large volumes of sludge that might require proper disposal or treatment before disposal. There will also be the potential of releasing residual coagulants or flocculants into the treated water, that may require additional treatment.[23]

RO requires high energy consumption due to its operating conditions requiring high pressure to overcome osmotic pressure. The RO method is also prone to membrane fouling, which reduces the filtration efficiency and requires frequent cleaning or replacement of membranes. This method requires high capital and operational costs associated with the use of RO systems.[24]

Thus, physical wastewater treatment methods play crucial roles in the removal of solid particles, contaminants, and dissolved substances from wastewater. The processes can be used as standalone processes or combined with other methods to achieve the desired wastewater quality after considering their limitations.

3.3.2 Chemical Treatment

Oxidation processes, such as advanced oxidation processes (AOPs), involve the generation of highly reactive oxidizing species to degrade and remove pollutants. These processes can effectively degrade recalcitrant organic compounds and remove color, odor, and taste from wastewater.[25] However, the effectiveness of oxidation processes can be influenced by the composition of the wastewater and the presence of interfering substances. Additionally, the high energy requirements and the generation of potentially harmful by-products are some of the limitations of oxidation processes.

Electrolysis involves the use of an electric current to drive chemical reactions in wastewater. Electrochemical treatment methods, such as electrocoagulation and electrooxidation, are commonly used in wastewater treatment. Electrolysis processes can effectively remove various pollutants, including organic compounds, heavy metals, and nutrients.[26] However, electrolysis processes can be energy-intensive and require careful control of operating conditions to achieve optimal treatment efficiency. The scalability and cost-effectiveness of electrolysis processes can also be limiting factors in their application.

Ozonation is a widely used chemical treatment process in wastewater treatment. Ozone (O_3) is a strong oxidizing agent that can effectively degrade and remove various pollutants, including organic compounds, microorganisms, and emerging contaminants.[27] Ozonation processes can be applied as standalone treatments or in combination with other processes, such as biological treatment or activated carbon adsorption, to enhance pollutant removal. However, the efficiency of ozonation can be influenced by factors such as pH, temperature, and the presence of certain compounds that can consume ozone. The cost of ozone generation and the potential formation of harmful by-products, such as bromate ions, are also considerations in the application of ozonation processes.[28]

Limitation of adapting oxidation for wastewater treatment is its stability and persistence of certain complex organic compounds that can make them resistant to oxidation, which requires longer treatment times or additional oxidation processes. During the oxidation process, there could be formation of harmful by-products, such as disinfection by-products, which will affect the water quality and regulatory compliance. When the process requires advanced oxidation processes (AOPs), the energy requirements will be high for it to fully utilize the ozone or other powerful oxidants. The effectiveness of this process can be influenced by the presence of interfering substances such as dissolved organic or inorganic matter in the wastewater, which will reduce its efficiency.[29]

Electrolysis processes are energy-intensive, especially when high current densities are required for efficient pollutant removal. The scalability and cost-effectiveness of electrolysis is one of the limiting factors, especially for large-scale wastewater treatment applications. Similar to oxidation, the effectiveness of electrolysis can vary depending on the presence of interfering substances. Electrolysis processes may require strict operating conditions, such as pH and electrode materials to achieve optimal results.[30]

Ozonation processes may require significant energy requirements for ozone generation, especially for large-scale applications. The efficiency of ozonation is influenced by the characteristics of wastewater, such as its pH, temperature, and the presence of certain compounds that can consume ozone. Similar to oxidation, ozonation might produce harmful by-products, such as bromate ions during the ozonation process that can affect the water quality and regulatory compliance.[27]

In summary, oxidation, electrolysis, and ozonation are chemical treatment processes that can effectively remove pollutants from wastewater. These processes have advantages in terms of pollutant degradation and removal. However, they also have limitations, including energy requirements, potential formation of harmful by-products, and the need for careful control of operating conditions. These limitations need to be considered when selecting and designing chemical treatment processes for wastewater treatment applications.

3.3.3 Biological Treatment

The modified contact-stabilization method is an aerobic biological treatment process that utilizes microorganisms to degrade organic pollutants in shrimp cracker industrial wastewater. The method involves the use of a modified contact-stabilization tank, where wastewater is aerated and mixed with activated sludge containing diverse microbial communities.[31] The microorganisms in the activated sludge metabolize and break down organic compounds present in the wastewater, converting them into carbon dioxide, water, and new microbial cells. This process helps to reduce the organic load and improve the quality of the treated wastewater.

The use of microorganisms in biological treatment processes offers several advantages. Microorganisms have diverse enzymatic activities that enable them to degrade a wide range of organic pollutants present in wastewater. They can efficiently remove organic matter, nutrients, and other contaminants, contributing to the overall purification of wastewater. Bacteria, in particular, are commonly used in biological treatment processes due to their abundance in sewage water and their ability to perform various enzymatic reactions. The presence of a rich phylogenetic and functional diversity of microorganisms in wastewater sludge enhances the treatment efficiency and resilience of the biological treatment system.

Enzymes also play a crucial role in biological treatment processes. They catalyze specific biochemical reactions, accelerating the degradation of organic pollutants and enhancing the efficiency of wastewater treatment. Enzymes, such as hydrolases and oxidoreductases, are commonly used in wastewater treatment due to their ability to break down complex organic compounds. Enzymatic processes can save time by facilitating the transport of substrates into cells, making the treatment processes more effective.[32]

However, there are limitations to the use of microorganisms and enzymes in biological treatment processes. The effectiveness of microbial degradation and enzymatic reactions can be influenced by various factors, including temperature, pH, oxygenation level, and the presence of inhibitors or toxic compounds. The presence of xenobiotic compounds in wastewater can reduce microbial communities, disrupt microbial diversity, and diminish the performance of biological treatment systems.[33] Additionally, the toxicity of industrial wastewater can limit the application of biological methods. Further research is needed to optimize the conditions for microbial growth and enzyme activity, improve the stability and efficiency of enzymatic processes, and address the challenges associated with the treatment of specific pollutants and complex wastewater matrices.

Overall, the use of microorganisms and enzymes in biological treatment processes offers a promising approach for the removal of pollutants in wastewater. These methods can contribute to the sustainable and efficient treatment of wastewater, reducing environmental pollution and protecting human health. Continued research and technological advancements are essential to further enhance the performance and applicability of biological treatment processes in wastewater treatment.

3.4 Advanced EDC Removal Technologies

Traditional water sources are under stress from pollution, overuse, and changing climate patterns, necessitating the exploration and implementation of innovative water treatment technologies. Advanced techniques for EDC removal still use the same concepts as conventional methods but adapted with improvements. EDC-containing wastewater should be treated effectively using eco-friendly technologies to reduce the impact on the environment and natural water bodies. Among these advancements, adsorption-based, membrane-based, and advanced oxidation technologies have risen to prominence as key solutions in securing a sustainable and reliable water supply.

3.4.1 Adsorption-based Technologies

Adsorption is a process by which molecules or particles adhere to the surface of a solid material, known as an adsorbent. In the context of EDC removal from wastewater, adsorption is an effective method because EDCs can bind to the surface of certain materials, reducing their concentration in the water. The commonly used adsorption-based technologies include the following.

Activated carbon: Activated carbon is a highly porous material with a large surface area, providing ample sites for EDC molecules to adsorb onto. It can be used in powdered or granular forms and can effectively remove a wide range of EDC, as well as other organic and inorganic contaminants.[34] *Biosorbents:* Biosorbents are natural materials or modified biomaterials that exhibit affinity for EDC molecules due to functional groups like hydroxyl, carboxyl, or amino groups. Examples include agricultural residues like rice husks, coconut shells, and even algae. Biosorbents are eco-friendly and can be a cost-effective option.[35] *Nanostructured adsorbents:* Nanomaterials like carbon nanotubes, graphene, and metalorganic frameworks (MOFs) have high surface areas and unique surface properties. These materials can offer enhanced adsorption capacity and selectivity due to their nanoscale features.[36] *Polymeric resins:* Synthetic polymers can be tailored to have specific functional groups that are selective for certain types of EDC. These resins can be used in fixed-bed columns for continuous EDC removal.[37]

Adsorption processes can be carried out by either photocatalysis or using polymer membrane based methods depending on several factors, including the nature of the EDC, the overall treatment goals, efficiency considerations, cost, and environmental impact. In many cases, a combination of methods might be optimal. For example, a preliminary step using polymer membrane-based adsorption could be followed by photocatalysis for further degradation. Ultimately, the choice depends on a comprehensive evaluation of the specific EDC and wastewater characteristics, treatment goals, cost considerations, and potential environmental impacts.

3.4.2 Membrane-based Technologies

Membrane processes involve the use of semi-permeable barriers to separate contaminants from water.[38] The membrane-based filtrations that are commonly used involve four main types: microfiltration (MF), ultrafiltration (UF), nanofiltration (NF), and RO. UF and NF membranes have different pore sizes that allow them to separate EDC based on size and charge. UF is effective for removing larger particles, while NF can remove smaller molecules and some divalent ions. RO: RO utilizes a dense membrane to separate solutes, including EDC, from water. It is particularly effective at removing ions and small molecules, achieving a high degree of purification.

Reverse osmosis, a membrane filtering technique, has proven to be highly efficient in eliminating dissolved micropollutants from drinking water systems.[39] During the procedure, the primary mechanism for the rejection of dissolved contaminants is through the utilization of tiny molecular weight cut-off (MWCO) and membrane pore widths, as well as the repulsion caused by the electrostatic charge between the dissolved contaminants and the membrane.[40] Moreover, the successful separation of pollutants occurs when the compound characteristics have low pK_a and K_{ow} values (<2), as well as when the solvent permeability of RO treatment is minimal.[41] The thin-film composite membrane (TFC) is widely recognised as the superior membrane in terms of its exceptional selectivity, resistance to compaction, and chemical stability in RO treatment, when compared to other membrane polymers like polyamide, polybenzimidazoline, and poly(piperazine-amide).[42]

RO membrane filtering has been found to successfully remove emerging compounds, including personal care products (PPCPs), insecticides, and BPAs.[43] According to the study conducted by Wang *et al.* (2018), it was shown that the majority of pharmaceuticals and PPCPs exhibiting an overall removal effectiveness over 95% in the final treated water. This high removal efficiency was particularly evident for PPCPs compounds present at concentrations below 10 μg L^{-1}.[44] Furthermore, Katibi *et al.* (2021) observed a high level of rejection (≥98%) for the separation of bisphenol A (BPA) by the use of polyamide-based RO membranes.[45] However, as compared to other methods of membrane filtration such as NF and UF, the primary disposal method for the concentrates (also known as brine) produced from RO treatment is by discharge into surface water bodies. The RO concentrates account for approximately 20% of the initial concentration of rejected pollutants.[46] One of the challenges encountered in RO is the effective management of brine, which is a by-product formed during the filtering process which has detrimental impacts on the environment.

The NF membrane technique has been acknowledged as a prospective approach for the removal of micropollutants, including hormones and pharmaceutical contaminants, from water and wastewater systems.[47] NF membranes operate by applying hydraulic pressure to drive water through nanopores, typically ranging from 0.2 nm to 0.4 nm. The adsorption of pollutants onto the membrane occurs through charge and size interactions, as shown by previous studies.[48] Furthermore, it has been observed that NF membranes possess the ability to selectively separate dissolved components based on their charge.[49] According to the cited source, monovalent ions and water have the ability to penetrate, but divalent and multivalent ions are unable to pass through. In practical applications, it is well recognised that NF membrane treatment has excellent performance in terms of producing high-quality effluents with low organic content. Additionally, the removal of bacteria and viruses may be achieved without the need for chemical disinfectants.[50]

In their study, Datel and Hrabankova (2020) observed that the selection of membrane adsorbents, including polysulfone, ceramic acetate, polyacrylonitrile, and polyethersulfone, as well as the distinct size exclusion and charge repulsion characteristics of different EDC compounds, may significantly influence the extent of removal, ranging from 10% to 99.9%.[51] Therefore, it is crucial to comprehend the processes of NF membranes that are responsible for the rejection of certain contaminant molecules. In a study conducted by Semião *et al.* (2013), it was shown that the pore radius of the active layer has a crucial role in the elimination of adsorbing pollutants, such as estrone and estradiol, in NF membranes.[52] According to Huang *et al.* (2013), the adsorption and retention of hormones such as estrone and estradiol by NF membranes are influenced by a combination of partitioning effects and internal surface area access. The removal of various EDCs from water remains a challenge when relying only on individual treatment techniques. In the practical use of NF membranes in water treatment plants (WTPs), several approaches including multiple or hybrid treatments have been suggested to address the limitations associated with NF membranes. As a result, NF membranes are frequently integrated with advanced oxidation processes (AOP), and the handling of concentrated residuals necessitates further processing.[53]

UF is a membrane separation method characterized by low pressure, wherein the pore diameters of the membranes used typically fall within the range 0.01–0.1 µm.[54] UF membranes have been found to be efficient in the removal of macromolecules and particles. However, the effectiveness of these membranes is greatly influenced by the composition of the membrane material.[55] UF membranes have garnered significant interest due to their wide range of applications in advanced treatment of secondary effluent. Nevertheless, the use of UF membranes for the elimination of EDCs exhibits reduced efficacy. The rejection of target EDCs by UF membranes has been shown to be much lower when compared to NF and RO procedures, as described in a study.[56]

A comparative analysis conducted by Patel *et al.* (2019) determined that the efficacy of removing certain medications (such as amoxicillin, naproxen, metoprolol, and phenacetin) by UF is somewhat favourable when compared to NF processes.[57] Furthermore, Ojajuni *et al.* (2015) discovered that hydrophobic adsorption and size exclusion are the primary processes responsible for retaining EDCs on NF membranes.[58] Conversely, UF membranes tend to retain mostly hydrophobic EDCs. As a result, the use of UF as a standalone method for the elimination of EDCs is seldom observed. The restricted use of ultrafiltration membranes can be related to the disparity between their molecular weight cutoff (MWCO) range, which typically falls in the range 10–100 kDa, and the molecular weight of the majority of micropollutants, which is typically less than 1 kDa.[59]

In their study, Huang *et al.* (2019) examined the application of ultrafiltration in treating effluents. The effluents were subjected to a cross-flow process with a 1 kDa membrane, resulting in two distinct phases: a colloidal phase ranging from 0.45 µm to 1 kDa, and a soluble phase with molecules smaller than 1 kDa.[60] The researchers conducted a comparative analysis of the estrogenic activity in relation to other processes, including coagulation sedimentation (CS), GAC adsorption, magnetic ion exchange resin (NDMP), and ozone processes. The EDCs with lower K_{ow} values exhibit greater removal efficacy when subjected to NDMP, but the ozone method effectively eliminates both the colloidal and soluble phases of EDC. In this context, it is noteworthy that UF possesses the capability to permit the permeation of low-molecular-weight organic solutes through its membrane. Consequently, UF can serve as a preliminary treatment method for RO due to its ability to prevent highly concentrated effluent from reaching the RO membrane.

The membrane bioreactor (MBR) is another membrane-based technology. MBRs combine biological treatment (microbial degradation) with membrane separation.[61] Microorganisms in the bioreactor break down organic matter, and membranes separate the treated water from the biomass, ensuring effective dye removal. MBRs have gained significant interest in the field of biological treatment due to their technical advancements and ability to produce high-quality treated effluents, surpassing the capabilities of conventional biological treatments such as activated sludge treatment.[62] The effectiveness of MBRs is primarily attributable to the retention of sludge on the membrane surface. This retention mechanism facilitates comprehensive microbial degradation and physical retention of all molecules.[63] According to Hai *et al.* (2018), the efficacy of MBR treatment relies on several aspects including sludge age, concentration, the presence of anoxic and anaerobic compartments, wastewater composition, operating temperature, pH, and conductivity.[64]

The study conducted by Park *et al.* (2017) found that the use of MBRs is more effective in removing pharmaceuticals owing to the delayed degradation process. This is attributed to the advanced age of the sludge in MBRs, which results in the formation of unique microbial communities when compared to those seen in activated sludge plants.[65] Furthermore, Arca-Ramos *et al.* (2018) conducted a study in which they investigated the efficacy of one-stage and two-stage enzymatic MBRs for the elimination of micropollutants from secondary effluents.[66] In practical applications, MBR treatments have exhibited efficiency in the elimination of EDCs. A significant proportion, ranging from 81% to 99%, of diverse categories of EDCs, such as medicines, pesticides, and hormones with phenolic and estrogenic properties, have been effectively eliminated using MBR treatment, as reported in many studies.[67–69] Furthermore, it has been observed that several pharmaceuticals and PPCPs, including salicylic acid and propylparaben, have been completely removed (100% elimination) by the use of MBR therapy. However, several drugs, such ofloxacin, sulfamethoxazole, lorazepam, and propranolol, exhibit inadequate degradation when subjected to MBR treatment.[70] Consequently, other methods have been explored to address issues related to membrane fouling and current density.[71] Furthermore, the rates of membrane fouling have been seen to result in decreased removal efficiency. In light of this, it is important to seek various alternatives in order to examine the optimal configuration for achieving high treatment performance in MBR systems.

3.4.3 Advanced Oxidation Processes

Advanced oxidation processes (AOPs) involve the generation of highly reactive oxidizing species to degrade organic pollutants, including EDC. Fenton's process involves the addition of hydrogen peroxide and a ferrous iron catalyst to the water. This generates hydroxyl radicals ($^{\bullet}OH$), which are highly reactive and capable of oxidizing and breaking down dye molecules into simpler, less harmful compounds.[72] Fenton procedures are low-cost treatment technologies that function by degrading pollutants produced by the oxidation reaction of hydroxyl radicals. In practise, hydroxyl radicals target organic pollutants, leading in the full breakdown of contaminants into CO_2, water, and inorganic salts as end products.[73] Hydroxyl radicals take electrons from existing pollutants to produce hydroxide anions and obtain a hydrogen atom to replace the atom lost during the process.[74] The performance of Fenton treatments, in particular, is substantially determined by reaction process parameters.

According to Mirzaei *et al.* (2017), the parameters of reaction concentration, catalyst type, pH, radiation intensity, water matrix, substrate salinity, feeding method, temperature, and reaction duration are important for the performance of Fenton treatment procedures.[75] These techniques are significantly engaged in the removal of EDCs in water and wastewater treatment systems in general. Sun *et al.* (2016) used Fenton oxidation methods to successfully remove E3, BPA, diethylstilbestrol, E2, and EE2 from cow dung effluent, with removal efficiencies of 84.9%, 99.5%, 99.1%, 97.8%, and 84.5%, respectively.[76]

Fe(III) is commonly employed as a catalytic agent in photo-Fenton treatment to form hydroxyl radicals *via* a reaction with H_2O_2 under ultraviolet (UV) light.[77] In this process (ideally under acidic conditions), hydroxyl complexes that absorb light in the UV/visible region, such as Fe(OH)2+ and Fe(OH)2 4+, undergo photoreduction to create hydroxyl radicals and Fe^{2+}.[78] To obtain the best performance, an acidic environment is preferred; nevertheless, neutralization steps are necessary prior to discharge.[79] Photo-Fenton methods are currently being utilized to successfully remove a wide range of hormonal, phenolic, pesticide, and PPCPs chemicals. Photo-Fenton treatment has been shown to successfully remove pharmaceutical chemicals, with removal efficiencies ranging from 95% to 100%.[80]

The efficiency attained for the elimination of antibiotics, such as ciprofloxacin, amoxicillin, sulfathiazole, and sulfamethazine, falls within the range 80–95%. After a duration of 90 minutes, the photo-Fenton process, when exposed to both UV and solar light, demonstrated a reduction in total coliforms and *E. coli*, as reported in ref. 81. Silva *et al.* (2021) made a noteworthy discovery regarding the effective mitigation of estrogenicity (E2) and seven endocrine disruptors (EDs) through the utilization of LED irradiation as a viable substitute for solar photo-Fenton when solar radiation is not accessible.[82] This finding has significant implications for the reduction of potential risks associated with the reuse or discharge of wastewater treatment plant (WWTP) effluents.

The electro-Fenton method is employed to augment the production of hydroxyl radicals by the amalgamation of Fenton and electrocoagulation procedures. In the given circumstances, it can be observed that the oxidizing capacity of hydrogen peroxide (H_2O_2) is enhanced by the utilization of electrical aid in the electro-Fenton process.[83] Typically, electro-Fenton treatments are conducted using two distinct configuration conditions. The first involves utilizing inert electrodes with high catalytic activity as the anode material, while introducing Fenton reagents into the reactor externally. The second configuration involves externally adding hydrogen peroxide and providing Fe^{2+} from sacrificial cast iron anodes.[84] In this procedure, the electrochemical generation of H_2O_2 is employed to enhance the breakdown of high-strength organic contaminants inside wastewater systems.[85]

Specifically, electro-Fenton has demonstrated a high level of efficacy in the removal of elevated quantities of EDC contaminants during the treatment of wastewater.

3.4.4 Photocatalytic Membrane Reactor

Photocatalysis is a sophisticated physicochemical process and environmentally friendly technology capable of degrading air and water pollutants when exposed to a light source.[86] Photocatalysis has the capability to use light energy in an efficient manner for the purpose of eliminating a diverse array of hazardous organic pollutants from various mediums.[87] These pollutants include, but are not limited to, oil,[88] phenolic compounds,[89] surfactants,[90] colouring materials,[91] pesticides,[92] and some inorganic contaminants.[93] Photocatalysis is a highly favoured method for the comprehensive treatment of wastewater due to its ability to efficiently inactivate various microorganisms (such as protozoa, bacteria, viruses, and spores) within a reasonable timeframe. This approach offers several advantages, including the reduced need for chemical substances, the generation of non-toxic by-products and a lower production of secondary pollutants, as well as energy conservation.[94]

When a specific type of semiconducting material, referred to as a "photocatalyst," is subjected to illumination, the electrons (e^-) that originally reside in the valence band (VB) become energetically excited and transition to the conduction band (CB). Simultaneously, vacancies known as holes (h^+) are generated at their original positions.[95] The presence of these cavities results in the photocatalyst exhibiting a pronounced inclination towards the degradation of organic micropollutants. The by-products generated during the process of photocatalysis, namely carbon dioxide (CO_2) and water (H_2O), have the potential to be utilized in the synthesis of a useful product with diverse applications. Nevertheless, the process of photocatalysis sometimes overlooks the selective synthesis of these molecules and the generation of harmful by-products resulting from inadequate destruction of pollutants.[95] Therefore, the assessment of toxicity in wastewater treatment systems continues to be a significant parameter for quality evaluation.

Several different semiconducting materials have been employed as photocatalysts. Titanium dioxide (TiO_2) is frequently cited in literature due to its advantageous properties such as its relatively affordable cost, larger band gap, ability to generate hydroxyl ions in water-based settings, and its capacity to oxidize contaminants under normal atmospheric circumstances.[96] On the other hand, titanium dioxide (TiO_2) exhibits lower efficacy in the absorption of solar radiation. Consequently, numerous investigations have sought to enhance its light-absorbing capabilities across the ultraviolet (UV) to visible spectrum by introducing metal dopants such as silver ions (Ag^+), iron ions (Fe^{3+}), and cobalt ions (Co^{3+}), as well as non-metal dopants including nitrogen (N), sulphur (S), fluorine (F), carbon (C), boron (B), and phosphorus (P). This modification strategy aims to overcome the limitations associated with the utilization of pure TiO_2. Doping introduces an electron sink within the catalyst, hence promoting efficient charge transfer and optimizing the bandgap. Consequently, the capacity for visible light absorption is improved.[97] Furthermore, the addition of a dopant enhances the selectivity potential towards CO_2 and H_2O.

Photocatalytic membrane reactors (PMRs) have garnered significant interest in research endeavours due to their ability to integrate filtration and photocatalysis processes. These reactors have been extensively studied for their efficacy in degrading organic pollutants inside continuous flow systems.[98] In the context of PMRs, the primary function of the photocatalyst is to destroy the pollutants present in the system. Simultaneously, the membrane serves a dual purpose: firstly, it acts as a barrier to prevent the loss of photocatalyst nanoparticles, thereby enhancing their retention or loading; secondly, it functions as a filter to remove macromolecular organic matter from the wastewater, so contributing to the purification of the resulting permeate. Furthermore, PMRs offer other advantages such as easy scalability, excellent energy efficiency, and adaptability. Furthermore, the use of inorganic nanoparticles has been shown to enhance both the mechanical and permeability characteristics of the membrane.[99] Hence, photocatalytic membrane reactors are widely acknowledged as a highly effective, environmentally friendly, and auspicious technology for sustainable industrial implementations. These applications encompass the conversion of carbon dioxide into solar fuel, the purification of water and air, as well as the electrolysis of water to generate hydrogen.[100]

In their study, Kumari *et al.* (2020) provided a complete analysis and evaluation of the performance of PMRs in the treatment of persistent organic pollutants (POPs) present in wastewater when subjected to light stimulation.[101] The findings of their research indicate that PMRs have the capability to offer a cost-effective solution for the remediation of POPs. POPs refer to a class of lipophilic compounds that are of anthropogenic origin and possess a high degree of resistance to degradation. As a result of their tenacity, they have achieved global dispersion in limited amounts. The presence of persistent organic pollutants has been linked to a variety of detrimental impacts on human health, such as compromised neurodevelopment, immune system dysfunction, and decreased reproductive function. Numerous persistent organic pollutants exhibit the capacity to interfere with the regular operation of the EDC.[102] In their study, Chen *et al.* (2022) conducted a comprehensive analysis on the combined impacts of photocatalysis and membrane treatment in PMRs.[103] The PMRs are typically constructed using two different methods, namely slurry or immobilized. The photocatalysts present in the former are dispersed throughout a liquid medium, facilitating enhanced interaction with organic molecules in the solution and resulting in heightened efficiency of the photocatalytic process. Nevertheless, it is observed that the photocatalysts have a tendency to accumulate on the surface of the membrane, resulting in fouling. This phenomenon leads to a drop in the flow rate and a consequent reduction in the photo-catalytic efficiency. Moreover, the subsequent retrieval and processing of the suspended catalysts for the purpose of reutilization pose significant difficulties. When the photocatalysts are integrated into the membrane matrix or applied as a coating on its surface in the latter configuration, they have the ability to break down organic pollutants. Additionally, this design helps to reduce fouling on the membrane surface, leading to an extended membrane lifespan and improved efficiency in the removal of contaminants during wastewater treatment. Furthermore, the issue of recycling and secondary pollution of nano-photocatalysts is well addressed due to the immobilization of the photocatalysts within the membrane matrix.[104] Nevertheless, it is commonly observed that the catalytic efficiency of photocatalysts tends to diminish following the process of immobilization. Certain photocatalysts exhibit unstable catalytic efficiency as a result of the leaching of metal ions and photo-corrosion. Hence, it is imperative to develop photocatalysts that exhibit exceptional stability and efficiency to ensure their suitability for prolonged use in PMRs.

3.5 Conclusions

The widespread utilization of several EDCs in various contexts has led to heightened environmental contamination. The chemicals penetrate biological tissues and cause health difficulties due to their unique characteristics. Studying source identification, long-term monitoring, exposure routes, and mechanisms of action is essential to comprehend the extent of pollution and associated health hazards. It is also crucial to take measures to reduce exposure and prevent adverse biological effects caused by these compounds. Environmental endocrine-disrupting compounds have been linked to a range of illness in both experimental animals and humans over the past decade.

Removing EDCs is a complicated and difficult problem that necessitates the application of several remediation methods. Physico-chemical procedures can effectively degrade these substances, but may be limited in terms of cost and efficiency for certain molecules. Biological methods provide sustainable and cost-efficient options, albeit they may necessitate extended treatment durations and meticulous supervision to guarantee their efficacy. Membranes show great potential for separating EDCs. Novel membranes are being created with distinct compositions and architectures to enhance the separation efficiency.

An increasing trend in membrane research is expected for the simultaneous filtering and degradation of EDCs. Nevertheless, certain concerns must be addressed before these membranes can be effectively implemented for practical purposes. There are concerns about photocatalytic membranes due to the high energy consumption of artificial light sources. Investigating the utilization of natural sunlight to facilitate photocatalytic reactions shows potential as an eco-friendly method. Thorough experiments are needed to assess the long-term durability of photocatalytic membranes under illumination and in the presence of reactive oxygen species.

List of Abbreviations

AOP Advanced oxidation processes
EDC Endocrine disruption compound
POP Persistent organic pollutant
BPA Bisphenol A
UF Ultrafiltration
NF Nanofiltration
RO Reverse osmosis

Acknowledgements

The authors would like to acknowledge the VC-High Impact Research Grant for financial support UNI/F02/VC-HIRG/85517/P12-02/2022. We also acknowledged Research, Innovation and Enterprise Centre (RIEC), Universiti Malaysia Sarawak (UNIMAS) for managing the research activity.

References

1. T. Svingen, *et al.*, Using alternative test methods to predict endocrine disruption and reproductive adverse outcomes: do we have enough knowledge?, *Environ. Pollut.*, 2022, **304**, 119242.
2. M. A. La Merrill, *et al.*, Consensus on the key characteristics of endocrine-disrupting chemicals as a basis for hazard identification, *Nat. Rev. Endocrinol.*, 2020, **16**, 45–57.
3. B. S. Karthikeyan, J. Ravichandran, S. R. Aparna and A. Samal, DEDuCT 2.0: An updated knowledgebase and an exploration of the current regulations and guidelines from the perspective of endocrine disrupting chemicals, *Chemosphere*, 2021, **267**, 128898.
4. K. N. Palansooriya, *et al.*, Occurrence of contaminants in drinking water sources and the potential of biochar for water quality improvement: A review, *Crit. Rev. Environ. Sci. Technol.*, 2020, **50**, 549–611.
5. D. Surana, J. Gupta, S. Sharma, S. Kumar and P. Ghosh, A review on advances in removal of endocrine disrupting compounds from aquatic matrices: Future perspectives on utilization of agri-waste based adsorbents, *Sci. Total Environ.*, 2022, **826**, 154129.
6. S. Kumar, *et al.*, Recent Advancement in Nanotechnology for the Treatment of Pharmaceutical Wastewater: Sources, Toxicity, and Remediation Technology, *Curr. Pollut. Rep.*, 2023, **9**, 110–142.
7. A. G. Mukherjee, *et al.*, Elimination of microplastics from the aquatic milieu: A dream to achieve, *Chemosphere*, 2022, **303**, 135232.
8. R. Kiyama and Y. Wada-Kiyama, Estrogenic endocrine disruptors: Molecular mechanisms of action, *Environ. Int.*, 2015, **83**, 11–40.
9. R. H. Waring and R. M. Harris, Endocrine disrupters: A human risk?, *Mol. Cell. Endocrinol.*, 2005, **244**, 2–9.

10. A. I. Farounbi and N. P. Ngqwala, Occurrence of selected endocrine disrupting compounds in the eastern cape province of South Africa, *Environ. Sci. Pollut. Res.*, 2020, **27**, 17268–17279.
11. A. C. Gore, *et al.*, EDC-2: The Endocrine Society's Second Scientific Statement on Endocrine-Disrupting Chemicals, *Endocr. Rev.*, 2015, **36**, E1–E150.
12. International Program on Chemical Safety, Global Assessment of the State-of-the-Science of Endocrine Disruptors, 2002, https://cir.nii.ac.jp/crid/1570291226263715968.
13. N. Vilahur, *et al.*, Prenatal exposure to mixtures of xenoestrogens and repetitive element DNA methylation changes in human placenta, *Environ. Int.*, 2014, **71**, 81–87.
14. R. Jeselsohn, *et al.*, Emergence of Constitutively Active Estrogen Receptor-α Mutations in Pretreated Advanced Estrogen Receptor–Positive Breast Cancer, *Clin. Cancer Res.*, 2014, **20**, 1757–1767.
15. T. M. Attina, *et al.*, Exposure to endocrine-disrupting chemicals in the USA: a population-based disease burden and cost analysis, *Lancet Diabetes Endocrinol.*, 2016, **4**, 996–1003.
16. S. Maletz, *et al.*, In vitro characterization of the effectiveness of enhanced sewage treatment processes to eliminate endocrine activity of hospital effluents, *Water Res.*, 2013, **47**, 1545–1557.
17. A. Das, *et al.*, A multidimensional study of wastewater treatment, *Int. J. Exp. Res. Rev.*, 2022, **28**, 30–37.
18. S. Dagar, S. K. Singh and M. K. Gupta, Integration of Pre-Treatment with UF/RO Membrane Process for Waste Water Recovery and Reuse in Agro-Based Pulp and Paper Industry, *Membranes*, 2023, **13**, 199.
19. D. Ramutshatsha-Makhwedzha, A. Mavhungu, M. L. Moropeng and R. Mbaya, Activated carbon derived from waste orange and lemon peels for the adsorption of methyl orange and methylene blue dyes from wastewater, *Heliyon*, 2022, **8**, e09930.
20. D. Dadebo and D. Obura, Removal of Acid Red 88 from an Aqueous Solution Using Kaolinite Clay by Adsorption Process, *East Afr. J. Eng.*, 2022, **5**, 57–71.
21. Y. Shiva Shankar, K. Ankur, P. Bhushan and D. Mohan, Utilization of Water Treatment Plant (WTP) Sludge for Pretreatment of Dye Wastewater Using Coagulation/Flocculation BT, in *Advances in Waste Management*, ed. A. S. Kalamdhad, J. Singh, and K. Dhamodharan, Springer, Singapore, 2019, pp. 107–121.
22. E. Rápó and S. Tonk, Factors Affecting Synthetic Dye Adsorption; Desorption Studies: A Review of Results from the Last Five Years (2017–2021), *Molecules*, 2021, **26**, 5419.
23. J. Bratby, Coagulation and Flocculation in Water and Wastewater Treatment, *Water Intell. Online*, 2016, **15**, 9781780407500.
24. K. P. M. Licona, *et al.*, Assessing potential of nanofiltration and reverse osmosis for removal of toxic pharmaceuticals from water, *J. Water Process Eng.*, 2018, **25**, 195–204.

25. R. Javaid, U. Y. Qazi, A. Ikhlaq, M. Zahid and A. Alazmi, Subcritical and supercritical water oxidation for dye decomposition, *J. Environ. Manage.*, 2021, **290**, 112605.
26. X. Huang, *et al.*, Degradation of organic dye wastewater by H2O2-enhanced aluminum carbon micro-electrolysis, *Environ. Sci. Pollut. Res.*, 2022, **29**, 72586–72597.
27. S. N. Malik, P. C. Ghosh, A. N. Vaidya and S. N. Mudliar, Hybrid ozonation process for industrial wastewater treatment: Principles and applications: A review, *J. Water Process Eng.*, 2020, **35**, 101193.
28. C. V. Rekhate and J. K. Srivastava, Recent advances in ozone-based advanced oxidation processes for treatment of wastewater- A review, *Chem. Eng. J. Adv.*, 2020, **3**, 100031.
29. D. B. Miklos, *et al.*, Evaluation of advanced oxidation processes for water and wastewater treatment – A critical review, *Water Res.*, 2018, **139**, 118–131.
30. R. Fu, P.-S. Zhang, Y.-X. Jiang, L. Sun and X.-H. Sun, Wastewater treatment by anodic oxidation in electrochemical advanced oxidation process: Advance in mechanism, direct and indirect oxidation detection methods, *Chemosphere*, 2023, **311**, 136993.
31. J. Jaafari, A. Bakhsh, H. Barzanouni, A. Younesi and N. Amir, Performance of modified one-stage Phoredox reactor with hydraulic up-flow in biological removal of phosphorus from municipal wastewater, *Desalin. Water Treat.*, 2019, **175**, 216–222.
32. P. Jain, A. G. Bhavani, P. Singh and M. B. Singh, Fundamental of Aerobic and Anaerobic Processes in Dye Wastewater BT, in *Biological Approaches in Dye-Containing Wastewater*, ed. A. Khadir, and S. S. Muthu, Springer, Singapore, 2022, vol. 1, pp. 39–55, DOI: 10.1007/978-981-19-0545-2_2.
33. A. Singh, *et al.*, Biological remediation technologies for dyes and heavy metals in wastewater treatment: New insight, *Bioresour. Technol.*, 2022, **343**, 126154.
34. M. F. M. A. Zamri, *et al.*, Treatment strategies for enhancing the removal of endocrine-disrupting chemicals in water and wastewater systems, *J. Water Process Eng.*, 2021, **41**, 102017.
35. M. A. Ahsan, *et al.*, Biosorption of bisphenol A and sulfamethoxazole from water using sulfonated coffee waste: Isotherm, kinetic and thermodynamic studies, *J. Environ. Chem. Eng.*, 2018, **6**, 6602–6611.
36. V. Aggarwal, S. Solanki and B. D. Malhotra, Applications of metal-organic framework-based bioelectrodes, *Chem. Sci.*, 2022, **13**, 8727–8743.
37. D. Azizi, *et al.*, A comprehensive review on current technologies for removal of endocrine disrupting chemicals from wastewaters, *Environ. Res.*, 2022, **207**, 112196.
38. N. Shehata, *et al.*, Membrane-based water and wastewater treatment technologies: Issues, current trends, challenges, and role in achieving sustainable development goals, and circular economy, *Chemosphere*, 2023, **320**, 137993.

39. K. K. Razman, *et al.*, Environmental performance of a photovoltaic brackish water reverse osmosis for a cleaner desalination process: A case study, *Sci. Total Environ.*, 2023, **896**, 165244.
40. N. Moriyama, H. Nagasawa, M. Kanezashi and T. Tsuru, Water permeation in gas and liquid phases through organosilica membranes: A unified theory of reverse osmosis, pervaporation, and vapor permeation, *Chem. Eng. Sci.*, 2022, **263**, 118083.
41. J. R. Werber, A. Deshmukh and M. Elimelech, The Critical Need for Increased Selectivity, Not Increased Water Permeability, for Desalination Membranes, *Environ. Sci. Technol. Lett.*, 2016, **3**, 112–120.
42. S.-T. Fan, *et al.*, Enhanced selectivity in thin film composite membrane for CO_2 capture through improvement to support layer, *Chem. Eng. J.*, 2023, **468**, 143645.
43. A. Matin, S. M. S. Jillani, U. Baig, I. Ihsanullah and K. Alhooshani, Removal of pharmaceutically active compounds from water sources using nanofiltration and reverse osmosis membranes: Comparison of removal efficiencies and in-depth analysis of rejection mechanisms, *J. Environ. Manage.*, 2023, **338**, 117682.
44. Y. Wang, *et al.*, A Novel Tracheobronchial Stent Loaded with 125I Seeds in Patients with Malignant Airway Obstruction Compared to a Conventional Stent: A Prospective Randomized Controlled Study, *EBioMedicine*, 2018, **33**, 269–275.
45. K. K. Katibi, *et al.*, Recent Advances in the Rejection of Endocrine-Disrupting Compounds from Water Using Membrane and Membrane Bioreactor Technologies: A Review, *Polymers*, 2021, **13**, 392.
46. M. Umar, F. Roddick and L. Fan, Recent Advancements in the Treatment of Municipal Wastewater Reverse Osmosis Concentrate—An Overview, *Crit. Rev. Environ. Sci. Technol.*, 2015, **45**, 193–248.
47. S. Wang, *et al.*, A review of advances in EDCs and PhACs removal by nanofiltration: Mechanisms, impact factors and the influence of organic matter, *Chem. Eng. J.*, 2021, **406**, 126722.
48. S. Valizadeh, *et al.*, Experimental and density functional theory studies of laminar double-oxidized graphene oxide nanofiltration membranes, *Chem. Eng. Res. Des.*, 2022, **188**, 590–606.
49. L. Yang, *et al.*, High-permeable graphene oxide/graphitic carbon nitride composite nanofiltration membrane for selective separation of dye and desalination, *J. Environ. Chem. Eng.*, 2023, **11**, 109306.
50. S. Ideno, *et al.*, Phenotypic characterization of cell culture-derived hepatitis E virus subjected to different chemical treatments: Application in virus removal via nanofiltration, *J. Virol. Methods*, 2021, **296**, 114244.
51. J. V. Datel and A. Hrabankova, Pharmaceuticals Load in the Svihov Water Reservoir (Czech Republic) and Impacts on Quality of Treated Drinking Water, *Water*, 2020, **12**, 1387.
52. A. J. C. Semião, M. Foucher and A. I. Schäfer, Removal of adsorbing estrogenic micropollutants by nanofiltration membranes: Part B—Modeldevelopment, *J. Membr. Sci.*, 2013, **431**, 257–266.

53. Z. Huang, *et al.*, Performance evaluation of integrated adsorption-nanofiltration system for emerging compounds removal: Exemplified by caffeine, diclofenac and octylphenol, *J. Environ. Manage.*, 2019, **231**, 121–128.
54. A. G. Fane, C. Y. Tang and R. Wang, Membrane Technology for Water: Microfiltration, Ultrafiltration, Nanofiltration, and Reverse Osmosis, in *Treatise on Water Science*, ed. P. Wilderer, Elsevier, 2011, pp. 301–335, DOI: 10.1016/B978-0-444-53199-5.00091-9.
55. C. M. Sánchez-Arévalo, M. C. Vincent-Vela, M.-J. Luján-Facundo and S. Álvarez-Blanco, Ultrafiltration with organic solvents: A review on achieved results, membrane materials and challenges to face, *Process Saf. Environ. Prot.*, 2023, **177**, 118–137.
56. S. Sarp, Nanofiltration and ultrafiltration of endocrine-disrupting compounds, in *Environmental Contaminants and Endocrine Health*, ed. O. Carnevali and G. Hardiman, Academic Press, 2023, ch. 4.2, pp. 329–339, DOI: 10.1016/B978-0-12-824464-7.00007-6.
57. M. Patel, *et al.*, Pharmaceuticals of Emerging Concern in Aquatic Systems: Chemistry, Occurrence, Effects, and Removal Methods, *Chem. Rev.*, 2019, **119**, 3510–3673.
58. O. Ojajuni, D. Saroj and G. Cavalli, Removal of organic micropollutants using membrane-assisted processes: a review of recent progress, *Environ. Technol. Rev.*, 2015, **4**, 17–37.
59. C. J. Yehl and A. L. Zydney, Characterization of dextran transport and molecular weight cutoff (MWCO) of large pore size hollow fiber ultrafiltration membranes, *J. Membr. Sci.*, 2021, **622**, 119025.
60. Y. Huang, *et al.*, Distribution of endocrine-disrupting chemicals in colloidal and soluble phases in municipal secondary effluents and their removal by different advanced treatment processes, *Chemosphere*, 2019, **219**, 730–739.
61. S. Samsami, M. Mohamadizaniani, M.-H. Sarrafzadeh, E. R. Rene and M. Firoozbahr, Recent advances in the treatment of dye-containing wastewater from textile industries: Overview and perspectives, *Process Saf. Environ. Prot.*, 2020, **143**, 138–163.
62. C. Chiemchaisri, *et al.*, Membrane bioreactor for municipal solid waste leachate treatment and organic micropollutant removals, in *Current Developments in Biotechnology and Bioengineering*, ed. X.-T. Bui, W. Guo, C. Chiemchaisri, and A. Pandey, Elsevier, 2023, ch. 10, pp. 221–237, DOI: 10.1016/B978-0-443-19180-0.00018-3.
63. G. Faridizad, E. Abdollahzadeh Sharghi and B. Bonakdarpour, The use of membrane bioreactors in high rate activated sludge processes: How and why sludge retention time affects membrane fouling, *J. Water Process Eng.*, 2022, **47**, 102807.
64. F. I. Hai, K. Yamamoto and C. H. Lee, *Membrane Biological Reactors: Theory, Modeling, Design, Management and Applications to Wastewater Reuse*, IWA Publishing, 2nd edn, 2018.
65. J. Park, *et al.*, Removal characteristics of pharmaceuticals and personal care products: Comparison between membrane bioreactor and various biological treatment processes, *Chemosphere*, 2017, **179**, 347–358.

66. A. Arca-Ramos, G. Eibes, G. Feijoo, J. M. Lema and M. T. Moreira, Enzymatic reactors for the removal of recalcitrant compounds in wastewater, *Biocatal. Biotransform.*, 2018, **36**, 195–215.
67. J. Boonnorat, C. Chiemchaisri, W. Chiemchaisri and K. Yamamoto, Removals of phenolic compounds and phthalic acid esters in landfill leachate by microbial sludge of two-stage membrane bioreactor, *J. Hazard. Mater.*, 2014, **277**, 93–101.
68. Q. Wang, *et al.*, Insights into effects of operating temperature on the removal of pharmaceuticals/pesticides/synthetic organic compounds by membrane bioreactor process, *Environ. Pollut.*, 2023, **335**, 122145.
69. K. Wang, *et al.*, Anaerobic membrane bioreactor for real antibiotic pharmaceutical wastewater treatment: Positive effect of fouling layer on antibiotics and antibiotic resistance genes removals, *J. Cleaner Prod.*, 2023, **409**, 137234.
70. I. Alkarim, F. W. F. Wong, M. S. Mohamed, M. Halim and A. B. Ariff, A review on the application of biofilm-based bioreactors in the removal of pharmaceutically active compounds (PhACs) from wastewater, *J. Environ. Chem. Eng.*, 2023, **11**, 110226.
71. Q. Mei, *et al.*, New insight into the irreversible membrane fouling in different pore-sized ultrafiltration ceramic membrane bioreactors (UCMBRs) for high-strength textile wastewater treatment, *Chemosphere*, 2023, **331**, 138773.
72. G. Elumalai, B. Sowmya, R. K. Rajan and V. Shanmugam, Experimental study of photo electro-Fenton method for the removal of Reactive Yellow 186: Influence of operational parameters, *Environ. Prog. Sustainable Energy*, 2023, **42**, e14061.
73. C. Hu, *et al.*, The combination of Fenton process and Phanerochaete chrysosporium for the removal of bisphenol A in river sediments: Mechanism related to extracellular enzyme, organic acid and iron, *Chem. Eng. J.*, 2018, **338**, 432–439.
74. P. Grassi, *et al.*, Investigation of the reaction pathway for degradation of emerging contaminant in water by photo-Fenton oxidation using fly ash as low-cost raw catalyst, *Int. J. Environ. Res.*, 2020, **14**, 427–438.
75. A. Mirzaei, Z. Chen, F. Haghighat and L. Yerushalmi, Removal of pharmaceuticals from water by homo/heterogonous Fenton-type processes – A review, *Chemosphere*, 2017, **174**, 665–688.
76. M. Sun, *et al.*, Using Fenton Oxidation to Simultaneously Remove Different Estrogens from Cow Manure, *Int. J. Environ. Res. Public Health*, 2016, **13**, 917.
77. L. Clarizia, D. Russo, I. Di Somma, R. Marotta and R. Andreozzi, Homogeneous photo-Fenton processes at near neutral pH: A review, *Appl. Catal., B*, 2017, **209**, 358–371.
78. T. X. H. Le, *et al.*, Correlation between degradation pathway and toxicity of acetaminophen and its by-products by using the electro-Fenton process in aqueous media, *Chemosphere*, 2017, **172**, 1–9.
79. A. V. Vorontsov, Advancing Fenton and photo-Fenton water treatment through the catalyst design, *J. Hazard. Mater.*, 2019, **372**, 103–112.

80. M. Raji, S. A. Mirbagheri, F. Ye and J. Dutta, Nano zero-valent iron on activated carbon cloth support as Fenton-like catalyst for efficient color and COD removal from melanoidin wastewater, *Chemosphere*, 2021, **263**, 127945.
81. J. A. L. Perini, A. L. Tonetti, C. Vidal, C. C. Montagner and R. F. P. Nogueira, Simultaneous degradation of ciprofloxacin, amoxicillin, sulfathiazole and sulfamethazine, and disinfection of hospital effluent after biological treatment via photo-Fenton process under ultraviolet germicidal irradiation, *Appl. Catal., B*, 2018, **224**, 761–771.
82. L. G. R. Silva, *et al.*, LED irradiated photo-Fenton for the removal of estrogenic activity and endocrine disruptors from wastewater treatment plant effluent, *Environ. Sci. Pollut. Res.*, 2021, **28**, 24067–24078.
83. X. Wang, J. Zhao, C. Song, X. Shi and H. Du, An Eco-friendly Iron Cathode Electro-Fenton System Coupled With a pH-Regulation Electrolysis Cell for p-nitrophenol Degradation, *Front. Chem.*, 2022, **9**, 837761.
84. N. Oturan and M. A. Oturan, Electro-Fenton Process: Background, New Developments, and Applications, in *Electrochemical Water and Wastewater Treatment*, ed. C. A. Martínez-Huitle, M. A. Rodrigo, and O. Scialdone, Elsevier, 2018, pp. 193–221, DOI: 10.1016/B978-0-12-813160-2.00008-0.
85. P. Chanikya, P. V. Nidheesh, D. Syam Babu, A. Gopinath and M. Suresh Kumar, Treatment of dyeing wastewater by combined sulfate radical based electrochemical advanced oxidation and electrocoagulation processes, *Sep. Purif. Technol.*, 2021, **254**, 117570.
86. M. B. Tahir, S. Farman and M. Sagir, Advances in Photocatalytic Materials for Waste Water Treatment Applications, in *Encyclopedia of Smart Materials*, ed. A.-G. Olabi, Elsevier, 2022, pp. 759–767, DOI: 10.1016/B978-0-12-815732-9.00032-2.
87. S. Ahmed, *et al.*, Emerging pollutants and their removal using visible-light responsive photocatalysis – A comprehensive review, *J. Environ. Chem. Eng.*, 2021, **9**, 106643.
88. X. Hu, *et al.*, A step closer to real practice: Integrated tandem photo catalysis-biofilm process towards degradation of crude oil, *J. Environ. Manage.*, 2023, **342**, 118357.
89. K. A. M. Said, *et al.*, Relationship of Magnetic Strength to Zinc Ferrite Migration in Fabricating Photocatalytic Membrane for Phenol Photodegradation, *J. Environ. Chem. Eng.*, 2021, **9**, 105923.
90. V. G. Dileep Kumar, *et al.*, Singlet oxygen driven enhanced photocatalytic degradation of 1,3,7-trimethylpurine-2,6-dione using surfactant mediated PVA-CuO nanocomposites: Combining physical adsorption and photocatalysis, *Chem. Eng. J.*, 2023, **462**, 142187.
91. X. Gao, *et al.*, Comprehensive removal of various dyes by thiourea modified chitosan/nano ZnS composite via enhanced photocatalysis: Performance and mechanism, *Int. J. Biol. Macromol.*, 2023, **247**, 125677.
92. P. Kajitvichyanukul, *et al.*, Challenges and effectiveness of nanotechnology-based photocatalysis for pesticides-contaminated water: A review, *Environ. Res.*, 2022, **212**, 113336.

93. Y.-C. Lin, A. Kurniawan, J. A. A. Valinton, C.-H. Chen and S.-H. Chuang, Reduced graphene oxide–MoO_3 composites via microwave-assisted synthesis for dual-functional photocatalysis of organic dyes and heavy metal cation under simulated sunlight irradiation, *J. Taiwan Inst. Chem. Eng.*, 2023, 105013.
94. J. Liu, S. Zhang, W. Wang and H. Zhang, Photoelectrocatalytic principles for meaningfully studying photocatalyst properties and photocatalysis processes: From fundamental theory to environmental applications, *J. Energy Chem.*, 2023, **86**, 84–117.
95. C. B. Mendive and D. W. Bahnemann, Mechanisms in Heterogeneous Photocatalysis, in *Encyclopedia of Materials: Science and Technology*, ed. K. H. J. Buschow, R. W. Cahn, M. C. Flemings, B. Ilschner, E. J. Kramer, S. Mahajan, and P. Veyssière, Elsevier, 2011, pp. 1–5, DOI: 10.1016/B978-0-08-043152-9.02273-9.
96. Z. Huang and H. Liu, Insights into the pathways, intermediates, influence factors and toxicological properties in the degradation of tetracycline by TiO_2-based photocatalysts, *J. Environ. Chem. Eng.*, 2023, **11**, 110587.
97. A. S. M. Nur, *et al.*, A review on the development of elemental and codoped TiO_2 photocatalysts for enhanced dye degradation under UV–vis irradiation, *J. Water Process Eng.*, 2022, **47**, 102728.
98. C. Yang, *et al.*, A novel vertical immobilized photocatalytic membrane reactor based on Bi_2WO_6-g-C_3N_4/PVDF for enhanced removal of atrazine, anti-fouling performance and long-term stability, *Chem. Eng. J.*, 2023, **471**, 144672.
99. M. J. Benotti, B. D. Stanford, E. C. Wert and S. A. Snyder, Evaluation of a photocatalytic reactor membrane pilot system for the removal of pharmaceuticals and endocrine disrupting compounds from water, *Water Res.*, 2009, **43**, 1513–1522.
100. C. N. Rani, S. Karthikeyan and S. Prince Arockia Doss, Photocatalytic ultrafiltration membrane reactors in water and wastewater treatment - A review, *Chem. Eng. Process.*, 2021, **165**, 108445.
101. P. Kumari, N. Bahadur and L. F. Dumée, Photo-catalytic membrane reactors for the remediation of persistent organic pollutants – A review, *Sep. Purif. Technol.*, 2020, **230**, 115878.
102. T. Damstra, Potential Effects of Certain Persistent Organic Pollutants and Endocrine Disrupting Chemicals on the Health of Children, *J. Toxicol., Clin. Toxicol.*, 2002, **40**, 457–465.
103. L. Chen, P. Xu and H. Wang, Photocatalytic membrane reactors for produced water treatment and reuse: Fundamentals, affecting factors, rational design, and evaluation metrics, *J. Hazard. Mater.*, 2022, **424**, 127493.
104. A. Joseph and A. Vijayanandan, Review on support materials used for immobilization of nano-photocatalysts for water treatment applications, *Inorg. Chim. Acta*, 2023, **545**, 121284.

CHAPTER 4

Ultrafiltration for Laundry Wastewater Treatment

NUR ALYAA SYFINA ZAKARIA, PEI SEAN GOH*, WOEI JYE LAU AND AHMAD FAUZI ISMAIL

Advanced Membrane Technology Research Centre, Faculty of Chemical and Energy Engineering, Universiti Teknologi Malaysia, 81310 Johor Bahru, Malaysia
E-mail: peisean@petroleum.utm.my

4.1 Introduction

The report by UNESCO for the UN-Water Conference[1] shows that 26% of the world's population, which is equivalent to almost 2 billion people, do not have clean water to drink, and 46% of the population do not have good sanitation. Nearly 1.7 billion people in cities will be experiencing water scarcity, and that number has been projected to double in the next 30 years. The worsening water scarcity issue has also increasingly affected the environment and ecosystem.[2] Bacteria and heavy metals such as mercury, phosphorus, and nitrogen are the most common pollutants found in these sources. The demand for water climbs by 1% every year resulting in the usage of alternative sources of water such as poorly drilled wells and springs, rainfall inflows from cities, and wastewater recycling.[3] The significant rise in pollutants to levels that pose a threat to the environment has caused concerns among society, leading to the emergence of several unexplored pollution issues. The presence of these pollutants diminishes the availability of freshwater resources

Chemistry in the Environment Series No. 14
Advanced Materials for Emerging Water Pollutant Removal
Edited by Pei Sean Goh, Devagi Kanakaraju, Anwar Iqbal and Ahmad Fauzi Ismail

Published by the Royal Society of Chemistry, www.rsc.org

derived from both surface water and groundwater sources. Furthermore, it is observed that aquifers are experiencing a decline in productivity and are becoming contaminated on a worldwide level. This can be attributed to various factors such as soil erosion, the infiltration of saltwater, and the presence of fertilizers, pesticides, detergents, and heavy metals.[4]

Wastewater can spread infections such as cholera, typhoid, dysentery, and other diarrheal diseases brought on by poor sanitation killing millions of people every year.[5] Recently, tremendous efforts have been made to tackle the water scarcity problem by identifying alternative and sustainable water resources, with wastewater reclamation being the most prevalent method of producing usable water from wastewater.[6] Physical processes such as screening, raking, aeration basins, and settling ponds are used to remove biosolids during pretreatment at water treatment plants,[7] which treat sewage and industrial wastewater. Non-settleable materials are converted to settleable solids, which are then trapped in final clarifiers, while bacterial microbes are encouraged to devour organic material by the addition of oxygen. The reintroduction of treated water to the environment and the refilling of aquifers are generally safe when using advanced treatment technologies.[8]

There has been a growing focus on water sources used for non-drinking purposes. This is especially true for sources that generate large quantities of water and are in decentralized systems where water collection and treatment occur on-site, such as laundry wastewater (LWW). LWW is generated from the utilization of various cleansing agents such as soap, soda, and deodorant to effectively eliminate grime, oil, and stains from soiled garments. Various procedures are carried out in the laundry, such as cleaning white garments, using bleach, bluing, and applying starch, as well as employing distinctive methods for handling materials like wool and silk.[9] LWW also can be classified as a form of greywater, comprising elevated levels of concentration of chemical compounds that are derived from detergent powders and bleaching agents, along with suspended particulate matter and, potentially, traces of oils, paints, and other substances.

On average, washing machine produces 50–200 litres of wastewater per cycle.[10,11] Institutional, residential, and industrial settings cumulatively account for an average LWW output of 190–2106 litres per day.[12,13] In addition to a high chemical oxygen demand (COD) value (1200–25 000 mg L^{-1}), the surfactants, total suspended particles, nutrients, pH, pathogens, salts, and phosphates in LWW are also relatively high.[14] Surfactants such as linear alkyl benzene sulfonates, sulfonates, olein sulfonate, sulphates, and buffers like sodium tripolyphosphate constitute the backbone of detergents.[15] Water treatment will also become necessary when surfactants reach concentrations exceeding 0.5 mg L^{-1}.[16]

It is essential to implement appropriate treatments of LWW to eliminate contaminants before its release into the surrounding environment. The purification of wastewater generated from laundry before its discharge into aquatic systems is an effective measure that greatly contributes to the preservation of the natural environment while simultaneously improving the overall health and welfare of communities.[17] LWW treatment is particularly difficult due to the high concentrations of oils, surfactants, suspended particles, and COD, as well as the probable presence of organic solvents and heavy metals. Having a high level of turbidity, suspended particles, phosphorous compounds, allergic fragrance compounds, and other things, LWW can include over 20 000 mg L^{-1} of COD which commonly has a basic pH.[18]

LWW can be treated in several ways, which can be further classified into (i) biological, such as aerobic, anaerobic, and biofilter or wetland; (ii) physical, such as membrane separation, adsorption, electrocoagulation, and others; and (iii) chemical systems such as advanced oxidation process, photocatalysis, and Fenton; and a combination of physicochemical and biological treatments.[19] The processes of adsorption, coagulation, chemical precipitation, flotation, and biological treatments are those that are utilized most frequently in the treatment of wastewater from laundries.[20] However, the treated wastewater is often of insufficient quality to be utilized in the laundry system.[21] Therefore, the integration of physical, chemical, and biological approaches is sometimes required to efficiently treat LWW to meet the reuse standard.

UF membrane, which has MWCO in the range 300–500 000 Da, is one of the popular membranes used for wastewater treatment in industry. Most commercially available UF membranes are made from cellulose acetate and polysulfone. Several research studies on the use of various polymer materials have been conducted to increase membrane performance,[22–24] or decrease membrane costs.[25] With the increasing importance of implementing a circular economy, recent attention has been put on the development of sustainable membranes that are derived from waste such as plastic bottles. Some research has been conducted on the use of polyethylene terephthalate (PET) as a polymer material for the fabrication of UF membranes.[26–28]

The good mechanical properties, high transparency, and exceptional gas-barrier resistance make PET a widely used material in the production of packaging for food and beverages. PET bottles are now often used as single-use packaging for beverages such as soft drinks, mineral water, teas, and juices. However, millions of PET bottles are discarded as garbage every day, posing a severe environmental hazard because of their inability to biodegrade, taking 400 years to decay naturally, but which are easily recyclable.[29] Upcycling PET waste into a valuable membrane material offers an attractive opportunity to reduce the need to dispose of them as waste. PET could be turned into membrane materials[30] in many ways, including making electro-spun fibres,[31–33] ion tracked-etched membranes,[30,34] support layers for hollow fibre membranes,[35,36] and track-UV membranes.[37]

4.2 Laundry Wastewater

According to Manouchehri and Kargari (2017), the sources of LWW can be categorized into three main types: home, industrial, and hospital laundry.[38] It is characterized by elevated levels of chemical compounds derived from detergent powders, bleaching agents, as well as suspended particulate matter, and potentially including substances such as oils and paints.[39] The term "greywater" refers to wastewater produced by domestic processes such as bathing, showering, doing laundry, and cooking, but not to black water from the toilet. The amount of greywater produced by domestic activities contributes 50–80% of the total usage of water; among this waste, laundry greywater contributes up to 33%.[40,41] LWW is produced as the outcome of the use of soap, soda, and detergent to remove grease, dirt, and starch from dirty garments, the procedures include white-work washing, bleaching, bluing, and starching, as well as additional processes for wool, silk, and other materials.[9]

Even though LWW from commercial and domestic sources generally differ in the exact composition, the major contaminants of LWW always include phosphate, salt, potassium, calcium, magnesium, surfactant, fats, oils, suspended particles as well as organic micro-pollutants (OMPs). These impurities are eliminated from fabrics and washing detergents.[42] In addition, phosphate-free detergents were launched in 1986 to reduce laundry greywater's environmental impact and prevent eutrophication in the water bodies that received these wastes.[43] However, laundry greywater still contains a significant number of OMPs that are highly persistent in the environment.[44]

The primary constituent of detergent is surfactants, which are predominantly used to eliminate dirt from fabric arising from food waste, bodily wastes, and the environment. Surfactants are the main elements present in all laundry detergents that function effectively by detaching the dirt and oils from fabric surfaces and subsequently binding them to water molecules. Due to their usage in the production of detergents and hygiene products, which are used for bathing and washing clothing, surfactants (surface active agents) represent the majority of OMPs in greywater. Surfactants comprise the byproducts of detergents that are amphoteric, cationic, anionic, and non-ionic. These classes also include cationic and anionic surfactants such as benzoalkonium chloride, *n*-hexadecyltrimethoxylates, isotridecanolethoxylates, methyl ester sulphonate, olefin sulphonate, alkyl benzene sulphonates, alkyl ether sulphonates, and ammonium chloride.[44] Given the high concentration of surfactants used in LWW, which necessitates quick and precise action in removing dirt and oils from fabrics, it is also important to note that the presence of surfactants can potentially reduce the effectiveness of wastewater treatment plants.[13]

As shown in Figure 4.1, the discharge of wastewater resulting from laundry activities has some harmful impacts on both the environment and human health.[45] Thus, the debate over the environmental impact of detergents has mostly focused on the discharge of industrial and domestic wastewater into receiving waterways. The inadequate breakdown of surfactants results in

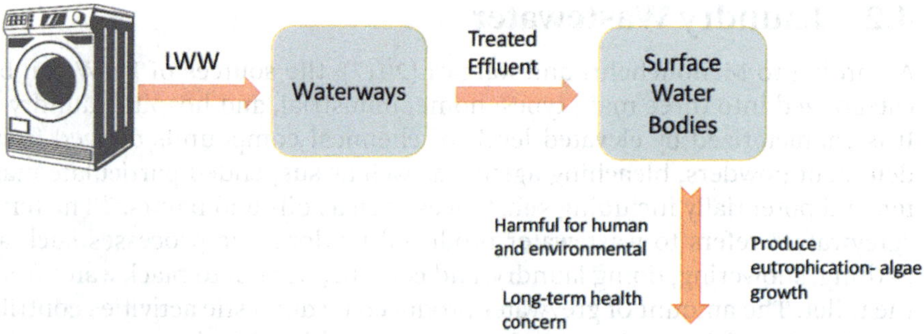

Figure 4.1 Human and environmental effects.

the formation of massive amounts of foam in streams and rivers near dams. Surfactants, in reality, can only be destroyed under specific conditions, such as when certain microbes are provided for their deterioration.[46] A layer of foam is typically produced on the surface of the water when there is a high concentration of detergent in the aquatic environment, which means that these substances slow down the rate at which oxygen passes through the water from the air, which leads to improper dissolved oxygen adsorption by aquatic organisms.[47,48] Builders such as phosphates are one of its components. Phosphate-based detergents enter the water due to excessive phosphate addition, producing eutrophication and, as a result, significant algae growth.[49,50] Finally, an increase in organic material dosages caused by algal death and degradation, result in a decrease in dissolved oxygen levels, which is an essential indicator of aquatic species mortality and biodiversity loss in the ecosystem. In addition to the aforementioned issues, phosphate and surfactant-polluted rivers are harmful to humans and aquatic life.[49,51]

1,4-Dioxane, a synthetic industrial chemical that is present in laundry detergent, is one of the components of a LWW which has been used extensively as a chemical stabilizer in other formulations.[52,53] It is most often connected with chlorinated solvents. It is also a waste product or contaminant in common household items like detergent. Although it is considered safe for use in tiny amounts, certain undesirable side effects might occur from being exposed to the chemical. Even more worrying is the ease with which this molecule can be absorbed by the skin, lungs, and digestive tract.[54,55] Because of the potential for long-term health concerns associated with even low levels of exposure, the U.S. Environmental Protection Agency (EPA) (2020), has categorized 1,4-dioxane as a Group B2 probable human carcinogen.[56] 1,4-Dioxane has been discovered in groundwater at locations around the United States, especially in Long Island's only source aquifer and in connection with former industrial and hazardous waste sites.[57] The toxicity and longevity of this compound in the environment provide a rising concern; once discharged, this developing contaminant is exceedingly costly to clean up from the environment, and 1,4-dioxane removal from drinking water supplies requires complex and expensive treatment technology.

4.3 Treatment Technologies for Laundry Wastewater

Treating laundry water before releasing it into aquatic systems contributes greatly to protecting the environment and improving the general well-being and living conditions of communities. Thus, effective LWW treatment is required to eliminate contaminants before it is released into the environment. Several studies have described the remediation of industrial LWW by a variety of methods, including biological, chemical, coagulation, floatation, and advanced oxidation processes.[39,58,59] The processes involved in treating LWW can be classified into three broad categories: physical, chemical, biological, as well as a hybrid of two or more of them.[13] Frequently, multiple methods are required to accomplish the intended water quality. Figure 4.2 summarizes the classifications of the treatment technology for LWW. The features of the treatment technologies as well as their limitations for LWW are listed in Table 4.1.

Based on Table 4.1, treatment technologies for LWW can be broadly divided into three methods: physical, biological, and chemical. The hybrid/integrated process is known as a more effective approach for treating LWW. For physical treatment, the process of physical purification involves the separation of particle matter and solids from industrial and sanitary effluents. The process involves removing various particles such as fabric, foliage, sand, and plastic components from wastewater. These particles have the potential to cause harm to treatment equipment and diminish the effectiveness of chemical and biological wastewater treatment methods. In order to tackle this matter, the utilization of practical and simple equipment is employed in the physical treatment procedures for both industrial and sanitary wastewater.[60]

There are a few physical treatments used to treat LWW such as membrane filtration, adsorption, and electrocoagulation. For example, several studies have highlighted the potential of reusing laundry wastewater that has been cleaned with a membrane for other purposes, such as flushing toilets. Lade and Gbagba (2017), employed a cellulose nitrate MF membrane to show that

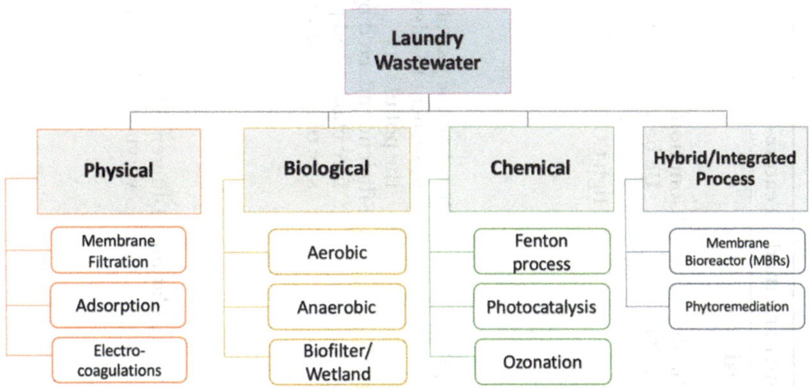

Figure 4.2 Summary of treatment technologies for LWW.

Table 4.1 Treatment technologies for LWW.

Treatment method	Features	Advantages	Drawbacks	References
Physical				
Membrane filtration	Commonly used MF and UF based on size exclusion mechanisms High COD rejection: 90%	Suitable for irrigation and flushing sanitary fixture-sUF move all suspended particulates	The permeate still contains organic compounds that are too high for bathwater reuse according to EU guidelines Operating at high pressures uses more energy Membrane fouling during filtering demands frequent cleaning to restore the permeate flux	38 and 61–63
Adsorption	Based on the use of adsorbent to adsorb the pollutants Efficiency affected by surface charge and pore size of adsorbents	An efficient, cost-effective, and simple method for cleaning up wastewater effluents that removes pollution Possible and effective adsorbents for removing heavy metals and radioactive substances from laundry wastewater, beyond the use of surfactants	The system is influenced by the surrounding environmental factors: initial pollutant concentration, adsorbent dosage, temperature, pH value, contact time, ionic strength, coexisting ions, and stirring speed	64 and 65
Electrocoagulation	Different electrodes show varying efficiencies	Easy to implement, cheap, environmentally friendly, flexible, effective, and generic – (eliminating practically all forms of pollution at once)	The electrodes form an overlaying film The sacrificial anode must be replaced regularly	66–68

Biological				
Aerobic	Involves adding a cultured type of bacteria to wastewater that is meant to break down certain pollutants Use air in the degradation process Balance pH and reduce the COD: 99.1% and BOD: 71%	Environmentally friendly process	Requires large amount of space to store biological waste Takes a longer time for different effluent treatments compared to chemical treatment	69–72
Anaerobic	Supplementation is another method to enhance anaerobic activity Do not use air in the degradation process Removal of COD: 89.8% and BOD: 94.0%	Low cost	Using commercial soap or its powdered variant led to an undesirable accumulation of volatile fatty acids in the effluent feeding to downstream processes due to the presence of additional compounds such as builders, enzymes, and optical	69–73
Biofilter or wetland	Constructed wetland consists of rocks in the lower part, soil in the middle, and plants in the upper part: divided into a vertical and horizontal flow Biofilter efficiency removal of TDS is 16.92% and TSS is 78.57%, phosphate is 83.28% and detergent is 93.1%	Convenient and inexpensive wastewater treatment	Low rejection for COD, BOD and TSS	74–78

(*continued*)

Table 4.1 (*Continued*)

Treatment method	Features	Advantages	Drawbacks	References
Chemical				
Fenton process	Capability to purify and disinfect wastewater streams containing persistent contaminants It can be done without a catalyst and with minimal work	Environmentally friendly Economically profitable	Electricity demand (UV lamps) increases costs	79 and 80
Photocatalysis	Around the photocatalyst, ultraviolet light creates highly oxidized hydroxyl radicals and oxygen anions The organic parts are then broken down into water and CO_2	Low cost High performance Minimal technological complexity	High concentrations of photocatalysis, create additional separation issues UV transmission is minimal because of the turbidity	79 and 81–85
Ozonation	Ozone is a strong oxidizing agent that breaks down organic materials and stops bacteria and viruses from doing their damage through a safe chemical reaction In any water purification process, ozone can be mixed with other oxidants, such as hydrogen peroxide, to make Advanced oxidation processes (AOPs)	Reduces costs: water, energy, and chemicals Treat laundry wastewater without generating any sort of secondary waste materials 13 times more dissolving power compared to oxygen	Limited mass transfer of ozone Short lifetime of the generated ozone and radical oxidative species (ROS) High energy consumption for ozone generation	86 and 87

Hybrid/Integrated process				
Membrane Bioreactor (MBRs)	Combinations of some membrane processes like microfiltration or ultrafiltration with a biological wastewater treatment process Implemented alongside additional pre-treatment processes Removal of turbidity: 99% and total solids (TS): 99%, while COD was between 70% and 99%	Benefits from compact form factor design Has excellent pollutant removal efficacy	Requires high energy consumption Increase operating cost due to membrane fouling Increase environmental footprint Challenges in the implementation of this technology	88–91
Phytoremediation	Using plants kiapu (*Pistia stratiotes*) and water hyacinth (*Eichhornia crassipes*) Followed by adsorption Remove COD: 77.5%, phosphate: 54.3%, and surfactants: 99.9% After centrifuging the feed greywater, microalgal phytoremediation using Chlorella variabilis Removal COD: 92.3%, BOD_5: 91.9%, TP: 85.6%, and TN: more than 97%	Microalgae – the oil byproduct possessed the potential to be harvested for biomass and biofuel production	Plant survival is affected by land toxicity and soil quality, hence plant-based remediation solutions cannot completely avoid groundwater contamination leaching Phytoremediation moves hazardous heavy metals, not removes them	88–91

the permeate was suitable for landscaping and flushing but not for other use because of its high biochemical oxygen demand (BOD) concentration.[62] Manouchehri and Kargari (2017) found 90% COD rejection in the permeate employing a different membrane-mixed cellulose ester (MCE) MF, although this method necessitated further dilution before reuse.[38] Using a tight UF membrane with a MWCO of 5 kDa, Ashfaq and Qiblawey (2018) studied the treatment of hospital laundry wastewater, finding that 90% COD removal was possible.[90] There is also a lot of interest in recycling closed-system greywater by membrane filtration, but there are still some obstacles to overcome. Using a rotating disk membrane with an NF270 membrane to treat detergent wastewater, Luo *et al.* (2012) produced good results, with a permeate flux of 350 L $m^{-2}h$ and 97% COD rejection.[92] In addition, greywater treatment on board sailing ships using MF has been studied, with 93% COD removal using the NF AC80 membrane.[93] However, there are downsides to membrane filtration that must be considered. These include membrane fouling, which necessitates frequent cleaning, and energy consumption at high pressures. To avoid these problems, a laundry wastewater treatment system that combines membrane filtration with other pre-treatment methods is ideal.

Adsorption on the other hand involves the use of adsorbents which contain active functional groups at the molecular level, which, through electrostatic interaction or coordination mechanisms, provide selectivity towards various surfactants. Due to ionic interactions, cationic resin adsorbents prefer anionic surfactants. To demonstrate the feasibility of targeting specific types of surfactant removal, anionic and cationic surfactant-preferring sorbents have been proposed by Schouten *et al.* (2007).[64] For example, Zhang *et al.* (2011) discovered that silica modified with amine-based polyelectrolytes enhanced the adsorption of the common anionic surfactant sodium dodecyl sulphate (SDS) found in cleaning and hygiene products.[94] Iron oxide powder (hematite and magnetite) has been suggested by Borghi *et al.* (2011) as an effective adsorbent for cationic surfactants, although a significant amount (17–51 g L^{-1}) was required to remove 90% of cationic surfactants in 10 minutes. The removal of pollutants from wastewater effluents through adsorption is regarded as an efficient and cost-effective method. Initial pollutant concentration, adsorbent dosage, temperature, pH-value, contact time, ionic strength, coexisting ions, and stirring speed all affect the efficiency and performance of adsorbents.[95] Understanding how adsorbents interact with various pollutants at the laboratory scale is essential for determining their potential application in laundry wastewater treatment and contribution to commercial-scale columns.

Using electrical charges from electrodes to mix ions or colloids with water makes electrocoagulation a way to treat wastewater. Jung *et al.* (2015) found that using granular activated carbon as a moving particle electrode can remove 98.6% of COD.[68] Meanwhile, most general regulations and guidelines for non-portable greywater reuse are met by this treatment. According to Janpoor *et al.* (2011), COD (93.2%), phosphorus (96.7%), detergent (93.5%), color (90.1%), and sediment were removed most effectively by an experiment using aluminum electrodes after 90 minutes of treatment.[67] Additionally, adding oxidizing agents like H_2O_2 during electrocoagulation helps make hydroxyl radicals and Fe^{2+} ions, which break down pollutants like SDS more easily.[96] Electrocoagulation has many benefits, such as being easy to use, inexpensive, not harmful to the environment, flexible, effective, and able to remove a wide range of pollution at once. However, regularly replacing the sacrificial anode is needed because the cathode passivates, and more work needs to be done to get electrocoagulation to the same level of understanding and reliability as other processes.

Furthermore, microorganisms, most commonly bacteria, are used in biological treatment procedures for the biochemical degradation of wastewater into stable end products. Sludges and other microbes proliferate, and some of the waste is broken down into carbon dioxide, water, and other elements. Based on the amount of dissolved oxygen present, biological treatment processes are typically classified as either aerobic or anaerobic. In aerobic bioreactors, degradation occurs.[97] Aeration in aerobic bioreactors enhances oxygen levels following anaerobic processes, which lack oxygen. Longer aeration in an aerobic bioreactor dissolves more oxygen in fermented waste. For example, Yang *et al.* (2012) demonstrated a 98.3% and 99.1% reduction in COD from a submerged membrane bioreactor for degrading artificial restaurant waste.[98] Dilafroza *et al.* (2013) reduced BOD by 71% in fluidized anaerobic bioreactor-based sewage treatment systems.[99] According to Chan *et al.* (2012), a thermophilic integrated anaerobic–aerobic bioreactor can reduce COD by up to 80%.[100] Meanwhile, wastewater treatment uses anaerobic bacterial metabolism in bioreactors.[101] An anaerobic bioreactor can achieve LAS degradation.[102] In biodegradation, bacteria alter the structure of the surfactant, causing it to lose its active surface.[103] Anaerobic bioreactors also lower detergent wastewater COD and BOD. According to Ji *et al.* (2020) two anaerobic membrane bioreactors were used to treat municipal wastewater, resulting in COD and BOD reductions of $89.8 \pm 1.8\%$ and $94.0 \pm 1.3\%$.[104]

Filtering contaminants with microorganism culture in inorganic media is simple and affordable with biofilters and constructed wetlands.[105] Biofilter media made of river rocks are affordable and simple to use. To absorb phosphate and surfactant, river rocks and granular activated carbon are utilized as biofilter media.[106] Constructed wetland (CW) wastewater treatment uses rocks, soil, and plants. Waste streams divide the system into vertical, horizontal, surface, and sub-surface flow. In this study, vertical surface flow allows wastewater to flow above the soil surface and fall to the bottom, potentially reducing pollutants. According to Choudhary and Kumar (2011), complex mechanisms in CW resulting from interaction between rocks, soil, and plants might lower total suspended solids (TSS), BOD, COD, nitrogen, pathogen, metals, and phosphate. The biofilter eliminates TDS and TSS efficiently (16.92% and 78.57%).[107] In the one-way ANOVA test, biofilter and CW outflows from the inlet had substantial variations in TDS and TSS. Three sample points had an average pH of 7.3–7.8, which is safe for aquatic life. Biofilter treatment removed most BOD, phosphate, and detergent after 4 weeks.

The addition of CW boosts removal rates in all parameters. Thus, the biofilter and CW systems removed 83.28% and 93.1% of phosphates and detergents, respectively.

Recently, biological treatment has been used ahead of chemical treatment. The term "treatment of wastewater by a process involving chemical treatment" currently encompasses both secondary and tertiary chemical treatments. According to Patil *et al.* (2020) when persistent pollutants are not eliminated by the biological process, advanced oxidation processes (AOPs) may be able to break them down.[15] The high hydroxyl radical generating efficiency of AOPs is linked to the complete mineralization of CO_2, water, and mineral compounds through the oxidation of several organic molecules.[108] The photo-Fenton process,[109] photocatalysis,[110] UV/H_2O_2,[111] and other AOPs are the most thoroughly studied.[112] These drawbacks include separation issues and high costs brought on by excessive electricity consumption or a large number of photocatalysts.[112] The capacity of photo-Fenton to disinfect and detoxify wastewater streams carrying persistent contaminants makes it an appealing choice over wastewater treatment.[80] It is profitable economically, does not require a catalyst, and is environmentally friendly.[113,114]

Hansson *et al.* (2012) also mentioned that photo-Fenton may work better than other AOPs when dealing with resistant contaminants. The $HO^•$ radical is created by the Fenton reaction, which combines Fe^{2+} with H_2O_2.[115] By employing UV light, the pace of the reaction is accelerated, and iron ions undergo cyclic reduction and oxygenation. The use of solar energy as a UV light source can significantly lower the cost of implementing this technology in practice.[116] The removal of persistent pollutants from laundry wastewater can be achieved by the solar photo-Fenton method, wherein highly oxidative hydroxyl radicals are the primary species involved in the process. The photocatalytic process works by using ultraviolet (UV) light to get rid of the water and oxygen in the area. This creates highly oxidized hydroxyl radicals and oxygen anions, which then break down the organic parts into water and CO_2. The whole process is safe and has no secondary pollution. Most of the time, the catalyst is attached to the surface of the ceramic support. The reaction efficiency is directly affected by the carrier qualities. Because washing machines need a lot of water, this project creates a new carrier structure, a new carrier material, and a reactor to make photocatalysis work better. When combined with the sterilizing effect of UV light, this makes the water purification process much more effective.[117]

By combining membrane techniques with others, such as coagulation, ion exchange, adsorption, or even other membrane processes, hybrid membrane processes can outperform individual technologies. It has been suggested that membrane bioreactors (MBR) be used alongside other pre-treatment procedures to not only increase the efficiency with which pollutants may be removed, but also to integrate them into a system that is capable of processing wastewater from beginning to end.[118] As the principal treatment unit, MBR is typically preceded by pre-treatment operations including physical screening and membrane filtration to remove suspended particles in the exit stream.[119] Nicolaidis and Vyrides (2014) tested a large-scale submerged aerobic MBR for 288 days to clean up effluent from an industrial laundry. A parabolic fine screen was used to filter out waste larger than 1.0 mm before the water reached the main MBR treatment unit which resulted in turbidity and total solids (TS) reduced by 99% and COD was removed by 70% to 99% in the treated wastewater, respectively.[120]

Siswoyo *et al.*, (2019) suggested treating LWW by combining phytoremediation with additional physicochemical treatments. To assist in phytoremediation and adsorption, they employed water hyacinth and kiapu plants. Up to 77.5% COD, 54.3% phosphate, and 99.9% surfactants could be eliminated by the combined phytoremediation system. In order to lessen suspended particles in greywater that the phytoremediation method was unable to remove, centrifugation was also suggested.[91] It was discovered by Oktor and Çelik (2019) that employing chlorophyll variabilis for microalgal phytoremediation reduced COD by 92.3%, BOD_5 by 91.9%, TP by 85.6%, and TN by greater than 97%.[121] Interestingly, harvesting the oil left over from microalgae could be used to produce biomass and biofuel.[122] Although MBR is the most widely used technique for treating laundry wastewater, membrane fouling makes it expensive to operate and consumes a lot of energy. Pre-treatment processes that selectively remove pollutants, such as membrane filtration, flotation, adsorption, and/or flocculation, are frequently used, although they have a negative impact on the environment and overall operating costs. Finally, recycled wastewater integration into existing infrastructures is challenging in developing nations, as they may lack the resources and developed infrastructures essential to support its implementation.[123]

Following the discussion of various technologies and processing methods, here it can be concluded as an attempt to reaffirm and recommend the most versatile fit process(es) in integrated configurations for the treatment of LWW. In most cases, the quality of treated wastewater produced by a single treatment stage is insufficient for it to be reused. However, recent developments in membrane manufacturing technologies have boosted the potential of single-stage membrane filtration as a practical method for treating laundry effluent. The use of UF membranes in a single-stage membrane filtration process has been shown to produce permeates that are appropriate for *in situ* recycling. Moreover, using a single-stage UF with a MWCO of 5 kDa also were able to treat hospital laundry wastewater to a quality where it could be reused in the United States, China, and Italy.

4.4 Principle and UF Membrane Configuration

According to Von M. Cheryan (1987), UF is a membrane technique used to remove the dissolved and colloidal material at low transmembrane pressure. Dissolved metal ions as low-molecular weights or hydrated ions could easily transmit UF membranes because their membranes have pores that are greater than dissolved metal ions. UF can be regarded as a technique that enables the purification, concentration, and fractionation of macromolecules or fine colloidal suspensions simultaneously.[124] UF is used in many areas these days, from the chemical industry to the food and drink industry. It is already widely used in many industrial settings, such as the food or pharmaceutical industries.[125] UF membrane technique's primary uses are in the purification and concentration of macromolecules, such as protein solutions in the food sector.[126–128] UF membranes have been constantly developed, and their applications have expanded to include chemical recovery, cell harvesting, dairy production, medicinal usage, wastewater reclamation, water treatment, and juice concentration.[129–133] The reasons why UF replaces conventional purification and disinfection processes are based on their simplicity and overall low cost due to low energy use, fewer control techniques, no or less reliance on process chemicals, moderate operating temperature, and high-quality treatment. Finally, the introduction of technology allows many businesses to become more environmentally friendly by allowing waste material recycling and resource recovery.[134,135]

Figure 4.3 displays a schematic illustration of the UF process. The process can be conceptualized as utilizing the pressure differential between the two sides of the membrane as the primary driving force. Driven by static pressure, the UF membrane serves as the filtering medium, allowing the solvent in the raw material liquid and the small-molecular-weight solute with a smaller pore diameter to permeate through the UF membrane from the high-pressure side to the low-pressure side. Conversely, the high-pressure side

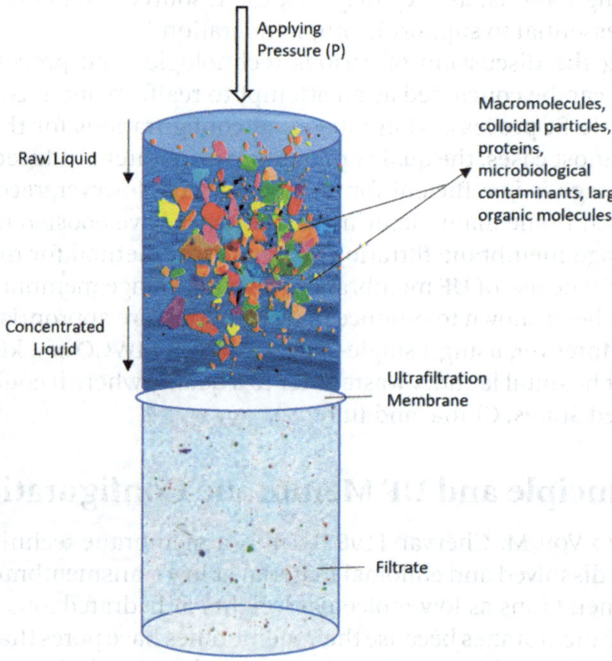

Figure 4.3 Schematic diagram illustrating the UF Process.

retains the large molecular weight solutes. In other words, the membrane selectively permits the passage of water, inorganic salts, and small molecular substances while inhibiting the permeation of macromolecules such as suspended solids, colloids, proteins, and microorganisms. This enables the membrane to effectively purify, separate, and concentrate the solution.[136,137]

The primary mechanism responsible for retaining particles in UF membranes is often recognized as size exclusion. Nevertheless, there are instances where the pore size of UF membranes exceeds that of both solvents and solute molecules. The chemical properties of the membrane surface, including the factors contributing to electrostatic effects, may be considered. In brief, the retention of the UF membrane can be attributed to three primary mechanisms: surface adsorption, pore retention, and elimination of mechanical pores on the membrane surface.[136]

4.5 UF Membrane Materials

UF membranes can be classified into two main categories: organic membranes and inorganic membranes. Organic membranes are produced through the utilization of synthetic polymers such as UF, polythene (PE), polytetrafluoroethylene (PTFE), polypropylene, and cellulose acetate are cited as examples.[138] Inorganic membranes encompass a range of materials, including ceramics, metals, zeolites, and silica. Membranes produced by various phase inversion procedures are predominantly employed for UF membranes. Various polymers, including poly(ethersulfone) (PES),[139] poly(sulfone) (PSf),[140] poly(vinylidenefluoride),[141] cellulose acetate (CA),[142] poly(acrylonitrile),[143] and their hybrids,[144] have been utilized in the fabrication of membranes with the desired composition and characteristics. Phase inversion may be described as a process of separating a homogenous polymer solution into distinct phases, resulting in the transformation of the solution from a liquid state to a solid state in a controlled manner.[145] The process under consideration can be characterized using many terminologies such as immersion precipitation, thermally induced phase separation, evaporation-induced phase separation, and vapor-induced phase separation.[146,147] The critical factors in the PI processes include solvent and non-solvent species, phase compositions, and temperatures of the polymer solution and the casting conditions, which must be properly taken into account and regulated[146,148–151] properties and separation performance of the membranes that are produced.

4.6 Challenges of UF

The issue of increased fouling in UF membranes during high flux operation continues to pose significant challenges in industrial applications.[152] Membrane fouling is a phenomenon that arises when particulate matter from the feed solution accumulates either on the surface of the membrane or within its porous structure.[153] The phenomenon of fouling in UF involves the accumulation of substances on the surface and/or within the pore structure of a membrane. The phenomenon of fouling has the potential to cause a permanent reduction in membrane permeability, in contrast to concentration polarization. Fouling exhibits the characteristic of being reversible. Numerous studies have made a distinction between reversible and irreversible fouling based on their respective resistance to cleaning.[154,155] Several cleaning techniques can eliminate fouling that is reversible in nature, but fouling that is irreversible persists despite these procedures. The phenomenon of hydraulically irreversible fouling is characterized by the inability to remove fouling deposits by hydraulic means. Chemical cleaning fails to eliminate chemically irreversible pollution.[129]

According to a prior investigation, the influence of pore size may be contingent upon the composition of the membrane, since hydrophilic membranes were found to exhibit reduced fouling compared to hydrophobic membranes possessing smaller pore sizes.[156] The reduction in overall flux can be significantly attributed to internal fouling resulting from in-pore adsorption, as the pore diameters of UF membranes are similar to those of several macromolecules. The distinction between adsorption and other forms of deposition lies in its thermodynamic nature, as opposed to deposition processes driven by body forces such as hydrodynamic forces imposed on colloids, which do not reach thermodynamic equilibrium.[157] Cake development occurs when particles build up on a membrane's surface, increasing permeate flow resistance. An initial cake layer of inert colloids at the membrane surface precludes direct interaction between active late foulants and the membrane surface. Next, is the process of gel formation, which occurs due to the consolidation of a layer of macromolecules that are highly concentrated in the immediate region of the membrane surface. This consolidation is driven by concentration polarization, as supported by previous studies.[158,159]

In practical applications, particularly in the context of UF processes involving complicated or mixed feeds, fouling phenomena can manifest as a combination of many forms operating concurrently. The significance of various fouling mechanisms is contingent upon the specific operating conditions, characteristics of the feed streams, and properties of the membrane.[160] In recent years, tremendous efforts have been put into the development of functional UF membranes through various modifications to improve their separation performance and antifouling properties.

4.7 UF Membrane Modification

Several modifications have been suggested to make UF membranes better at removing biofouling and fouling, and improving permeate flux and rejection. The common techniques used for UF membrane modification are shown in Figure 4.4.

For physical modifications, there are two main techniques involved which are membrane surface coating and blending technique. Coating the surface of a membrane is a simple, low-cost, and eco-friendly approach to surface

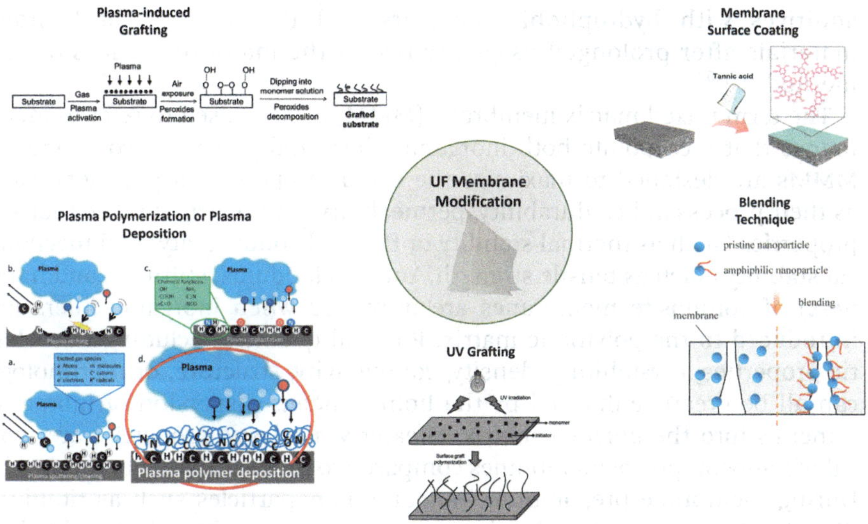

Figure 4.4 Illustration of the common types of membrane modification.

modification.[161] The goal of this method is to strengthen surface properties with minimal impact on the underlying structure. The most significant drawback of this technique is the instability of the coating layer. As a result, the performance and durability of surface coatings made with strong covalent bonding at the substrate-coating interface[162] are greatly improved. To further prevent the coated layer from penetrating the membrane pores, large molecular weight polymers are utilized.[163] For example, polydopamine coating which is a biopolymer, exhibits a remarkable adhesive behaviour similar to that of mussels. It undergoes self-polymerization in alkaline conditions (with a pH range of about 7.5 to 8.5) utilizing oxygen as an oxidizing agent. Consequently, it forms a thin layer on various substrates.[164,165] The application of a polydopamine coating results in the enhancement of surface hydrophilicity and the provision of anti-organic/fouling characteristics. In addition, it should be noted that polydopamine exhibits a variety of functional groups that can engage with a diverse array of molecules. This characteristic renders it a significant framework for the creation of covalently attached functional layers on a substrate.[166]

For the blending technique, blending polymers involves physically incorporating many compounds into a single casting solution while maintaining the same solvent. It is common practice to use additives in the phase separation process, such as polymeric compounds or inorganic nanofillers such as silica, silver, aluminum oxide, and titanium dioxide, to modify membrane characteristics. This is the simplest strategy to increase membrane hydrophilicity and decrease fouling, as most of these additions are hydrophilic in nature. The incompatibility of hydrophilic additives with hydrophobic polymers and the leaching of blended materials after prolonged usage are two of the major drawbacks of this method.[167,168]

The term mixed matrix membrane (MMM) is often used to refer to membranes that incorporate both inorganic fillers and polymeric components. MMMs are designed to maximize the useful properties of polymers, such as their processability, durability, permeability, and selectivity. The thermal properties (such as thermal stability or thermal conductivity) and mechanical stability (such as tensile strength, Young's modulus, stiffness, and hardness) of composite membranes are improved when inorganic fillers are introduced to the polymeric matrix. Physical qualities including viscoelastic properties, crystallinity, density, conductivity, structure, or morphology can all be greatly enhanced by the homogeneous dispersion of inorganic minerals into the polymer matrix. Greater viscoelastic properties are typical of nanocomposite membranes compared to those of pure membranes.[169] During membrane preparation, inorganic nanoparticles such as titanium dioxide, graphene oxide, alumina, silver, and copper have been added to polymer solutions to enhance membrane performance and properties. TiO_2 is a popular material due to its excellent chemical and thermal stability, minimal toxicity, photocatalytic activity, superhydrophilicity, and self-cleaning ability.

Anvari *et al.* (2019) used phase inversion to make UF MMM by adding TiO_2 nanoparticles to the PVDF/PAN casting solution. Thus, blended membranes demonstrated increased hydrophilicity, enhanced pure water flux, and improved antifouling properties.[170] Due to their comprehensive spectrum of antibacterial properties and low cytotoxicity,[25] Ag NPs have garnered a great deal of attention. They function as a local source of Ag^+ ions, preventing bacterial colonization and decreasing solute adhesion on membrane surfaces. The incorporation of inorganic materials has two disadvantages. First, nanoparticles are prone to aggregation during dope preparation, and second, there is a risk of NP release into the environment.[171] Porous materials such as mesoporous silica nanoparticles, nano-mineral materials, and metal–organic frameworks (MOFs) have been proposed as supports to control nanoparticle stability to resolve these issues.

Meanwhile, for blending organic materials, PVP and polyethylene glycol (PEG) are two examples of hydrophilic polymer additives used to improve membrane function and manufacturing. PVP's hydrophilicity makes it a pore-forming agent that accelerates phase inversion and creates a homogeneous, porous structure in the shape of fingers, which increases membrane flux. Increased PVP concentration, on the other hand, can inhibit macrovoid formation, hence decreasing water permeability. Hydrophilic additives like PEG are promising because they improve the permeability and pore development of polymeric membranes. PEG, which comes in a range of molecular weights, decreases the casting solution's thermodynamic stability, leading to the formation of porous structures resembling fingers. The membrane's morphology might shift from finger-like pores to spheres or ellipsoids with diminished mechanical properties if the molecular weight of PEG in casting solutions is increased from 200 to 10 000 Da.[172]

Surface grafting is one of the techniques used in chemical modification. Several techniques can be used to activate the membrane surface for grafting such as: (i) plasma-induced grafting, (ii) plasma functionalization or plasma activation, (iii) plasma polymerization or plasma deposition, and (iv) UV grafting. The addition of hydrophilic molecules, functional moieties, or electrostatically charged groups to the membrane surface is referred to as surface grafting. Surface grafting produces permanent surface modifications by covalently bonding grafted chains to the membrane.[163,173] This technique offers several benefits, including a simple modification procedure that permits relatively higher chemical stability and precise spatial control of grafting onto the desired surface.[174] However, it has the drawback of necessitating energy-intensive processes, resulting in an increase in membrane costs and scaling difficulties.

Plasma-induced grafting is plasma that is commonly regarded as the fourth state of matter, characterized by a partly ionized gas with quasi-neutral electrical properties. The generation of plasma commonly occurs by the application of microwaves or radio frequency waves to excite gases into higher energy states.[175] Upon exposure to a polymeric substance, various functional groups can be generated on its surface, so enabling future grafting or cross-linking events. UV grafting (photochemically-initiated graft polymerization) is widely used for modifying the surface of membranes. Using this approach, functional groups can be added with minimal impact on bulk characteristics. Without the need for additional photoinitiators, polyarylsulfone membranes can generate active sites or free radicals when exposed to light. When UV light is absorbed by the phenoxyl-phenyl chromophores in the backbone of PES polymeric chains, two radical positions (aryl and sulfonyl) are formed. Moreover, increased water permeability and decreased reverse salt diffusion have been achieved by modifying polyacrylic-polyethersulfone nanofiltration membranes *via* UV photografting. Surface roughness, hydrophilicity, fouling attachment, and water flux are all improved in functionalized membranes. However, the grafting process must be initiated by adding a photoinitiator or photosensitizer to other polymeric membranes, a process known as photoactivation.

4.8 UF for Laundry Wastewater Treatment

Recent advances in membrane separation technology have made it possible to treat groundwater and industrial effluents effectively. Stringent regulations made by environmental agencies also accounted for the cost of this membrane. The membrane separation process has several advantages over traditional methods. These include meeting stricter regulations, decreasing the negative effects of effluents on the environment, and using less space.[67,176] Table 4.2 lists previous studies that used various types of material for UF membranes and its characteristics for LWW.

Table 4.2 UF membranes and their characteristics for LWW treatment.

UF membrane	Remarks	LWW	Performance		References
			Flux	Rejection/Removal	
Hollow fibre ceramic membrane – zirconia, ZrO_2		Real laundry, industrial tent laundering in Hvalsø Teltudlejning ApS, Hvalsø, Denmark	Flux: 55 L m^{-2} h^{-1}	TSS: 95.9% VSS: 100% COD: 83.8%	177
Hollow fibre – poly(ether sulfone)	Batch 2: employed in the combined coagulation/flocculation/sedimentation and membrane separation UF processes to determine the best transmembrane pressure	Real laundry, industrial laundry (west of Parana, Brazil)	MSP parameters Transmembrane pressures (batch 2) Flux: 12.5 L m^{-2} h^{-1} Fouling: 76.7%	MSP parameters Transmembrane pressures (batch 2) TOC: 56.6% COD: 92% TN: 40.1% TDS: 50.5%	178
Hollow fibre – multichannel ceramic INSIDE CéRAM membrane with a MWCO 150 kDa	Application of a three-step system combining hybrid biological treatment followed by advanced UV/O_3 oxidation with *in situ* generated O_3 and membrane separation UF	Real laundry, industrial laundry Albatros Sp. z o. o. Sp. K. (branch: Nowe Czarnowo, Poland)	Flux: 102, 207, and 308 L m^{-2} h^{-1}	COD: 95% BOD_5: 92% TOC: 91% P_{total}: 29% N_{total}: 16%	21
Flat sheet – commercial TiO_2 ceramic membrane	Real laundry, University student dormitory in the school, South KoreaSynthetic laundry, SDBS, ammonium sulphate, magnesium sulfate heptahydrate (MgO_4S, 98% w/w), meat/beef extract powder (for microbiology), potassium phosphate monobasic (H_2KO_4P, 99% w/w), urea (CH_4N_2O, 98% w/w) and Bacto™ peptone		Synthetic LWW Flux: 2.5 L m^{-2} h^{-1}	TOC retention Synthetic: ~26.4% SDBS: ~21.6% Real LWW: ~9.4%	152
Flat sheet – polyethersulfone (PES) with MWCO 5 kDa	Real laundry, Hospital laundry, Qatar	Initial: 41.61 L m^{-2} h^{-1} Final: 37.82 L m^{-2} h^{-1}		TSS: 92.71% COD: 90% Total chlorine: 96%	90

Flat sheet – polyether-sulfone (PES) with 5 kDa MWCO	Samples were extracted from the pre-wash, wash and first rinse cycle of both the CZWP and UZWP programs	Real laundry, Doha hospital laundry	Flux: 29–42 L m^{-2} h^{-1}	Contaminated zone wash program (CZWP) TSS: 93% Total Cl: 96% COD: 90% Uncontaminated zone wash program (UZWP) TSS: 96% Total Cl: 100% COD: 94%	179
Flat sheet – polyacrylonitrile (PAN) with 75 kDa			Flux: 72–100 L m^{-2} h^{-1}	Contaminated zone wash program (CZWP) TSS: 89% Total Cl: 65% COD: 45% Uncontaminated zone wash program (UZWP) TSS: 88% Total Cl: 71% COD: 48%	179
Flat sheet – commercial polyethersulfone (PES) with 10 kDa	Membrane modified with hydrophilic polyvinylpyrollidone (PVP)	Real laundry, Laundry centre, National Institute of Technology campus, Tiruchirappalli, Tamilnadu, India	Flux: 55.2 L m^{-2} h^{-1}	COD: 88% TDS: 82%	39
UF on flat membranes: PVDF	UF on flat membranes is a post-treatment from Granular Activated Carbon (GAC) filtration	Real laundry, industrial laundry LIT S.r.l. (Turin, Italy)		COD: 87% TSS: 98%	180

According to Table 4.2, over the past 10 years, UF membrane has been used as a choice for LWW treatment. The UF membranes can be classified into two major configurations, *i.e.* hollow fibre and flat sheet membrane. In a study conducted by Luogo *et al.* (2022), a zirconia, ZrO_2 hollow fibre ceramic membrane was employed for the treatment of wastewater from a washing machine used for industrial tent cleaning.[177] Model microplastics (*i.e.*, nylon fibre) were added to deionized water, and the water was treated. The real wastewater from one wash cycle of PVC-fabric industrial tents was also treated. Through the use of 80 μm nylon fibers in synthetic feed filtration, the study was able to determine a critical flux value that indicated the pore-blocking effects of microplastics. With higher transmembrane pressure cycles, the UF membrane did not exhibit a significant flux drop, resulting in a 100% fiber removal rate. The actual wastewater from a tent laundry outlet had a 60 minute backflush interval and a critical flux value of 55 L m^{-2} h^{-1}. The permeability dropped by almost 37% during 4 days of continuous filtration, suggesting improved performance in real applications with less cleaning in-place requirements and longer operation times. Better water quality was achieved with nearly a 100% removal rate in all parameters after the UF membrane was filtered for an extended period of time. The UF ceramic membranes can also be employed in wastewater treatment systems of industrial laundries as polishing steps to get rid of microplastics before it is reused or released into municipal wastewater streams.

Nascimento *et al.* (2019) developed a hollow fibre made from poly(ethersulfone).[178] The efficacy of a combination of coagulation, flocculation, sedimentation, and membrane separation UF processes was compared to determine the best transmembrane pressure for treating laundry wastewater in relation to water quality physicochemical characteristics. The flux across UF membranes decreased rapidly in the first 10 minutes of filtration and then gradually decreased until stabilization, as seen by the permeation curves. The interaction between the membrane material and the other components of wastewater causes the fouling process, which leads to this reduction. For the UF membrane, the permeation flux (76.7%) was shown to grow linearly with the transmembrane pressure. The average permeability of UF membranes was 9.99 L h^{-1} m^{-2} bar^{-1}. The largest permeation fluxes were seen at a transmembrane pressure of 1.4 bar, suggesting that raising this pressure increases the driving force, hence facilitating the passage of liquid through the membrane's pores. As transmembrane pressure increased, the UF membrane's fouling values increased. In fact, higher flux decline values occur as suspended material in wastewater is decomposed on the membrane's surface during filtration. Moreover, maximum values for UF removal efficiency reached 92% as filtration pressure was elevated. Total organic carbon (TOC) removal efficiency utilizing UF membranes was 56.6%, while total nitrogen removal efficiency was 40.1%. The effluent composition and operating pressure are important parameters that govern the efficiency of the process.

Mozia *et al.* (2016) treated the effluent from a commercial laundry using a three-stage technique.[21] This was accomplished by UF membrane separation, enhanced UV/O_3 oxidation, *in situ* produced O_3, and a hybrid biological therapy. In UF, a ceramic INSIDE CéRAM multichannel membrane of 150 kDa was used under conditions of 1–3 bar transmembrane pressure and 4.5 m s^{-1} feed cross-flow velocity. Pure water flux (PWF-UF) was 102, 207, and 308 L m^{-2} h^{-1} for TMP of 1, 2, and 3 bar, respectively, across the applied membrane. In addition, TOC concentration was decreased by almost 90% thanks to the utilization of a hybrid moving bed biofilm reactor (HMBBR). The HMBBR effluent still had organic pollutants, high quantities of inorganic ions, and significant turbidity (8.2 NTU), therefore it needed to be treated further before it could be reused in the wash as pretreating the water with UV and O_3 before using UF was found to be an effective approach to reduce membrane fouling.

Kim and Park (2022) developed a study on commercial TiO_2 ceramic membrane to manage wastewater from laundry by using different cleaning techniques and analyzing fouling behavior and water flux recovery.[152] It was discovered that in synthetic laundry wastewater, lower applied pressure significantly reduced fouling resistance, increased TOC retention, and caused a flux decline. Cleaning techniques, such as those based on alkaline materials, enhanced the recovery of water flux because these methods were able to successfully separate the organic foulants affixed to the ceramic UF membrane. The study also discovered that the efficacy of cleaning strategies in terms of flux recovery and TOC retention is increased when chemical cleaning methods consisting of alkaline solutions followed by acidic solutions are used. Compared to synthetic wastewater, real LWW has a narrower flow decrease curve due to its higher particulate matter content and lower organic content. Organic materials were less successfully retained by the ceramic UF membrane than suspended particles. This research is the first to look into the fouling and flux behavior of ceramic UF membranes for treating LWW, and it shows that using them in combination with integrated techniques, such as RO membranes, could greatly increase retention rates.

LWW was treated with single-step membrane technology, specifically UF membranes with a reduced MWCO of 5 kDa by Ashfaq and Qiblawey (2018).[90] Most contaminants, including COD, surfactants, and suspended particles, can be eliminated with a single phase of treatment, and the permeate flux can still be maintained. According to the many wastewater reuse standards taken into consideration for the study, thus the treated water was found to comply. In terms of percentage rejection, it was discovered that the UF (5 kDa) membrane alone was unable to decrease wastewater conductivity, *i.e.* the dissolved solids rejection was minimal, as UF is not designed to remove mono and divalent ions. However, 90–100% of the suspended particles in the laundry effluent were rejected. Meanwhile, in terms of membrane flux, it was discovered that the flux initially dropped drastically when the experiment began, but then progressively stabilized over time. A similar pattern was seen while treating textile wastewater with a 10 kDa UF membrane.[181] This is known as the pseudo-steady state.[182] The pseudo-steady state occurs when the rate of foulant attachment on the membrane surface is in equilibrium with the rate of foulant removal from the membrane.

Ashfaq *et al.* (2017) applied two stages of UF treatment. A UF membrane with MWCO of 5 kDa was used for the first treatment stage and a UF membrane with a MWCO of 75 kDa was used for the second stage.[179] When exposed to raw laundry effluent, the 5 and 75 kDa UF membranes yielded flux values of 29–42 L m^{-2} h^{-1} and 72–100 L m^{-2} h^{-1}, respectively. 93%–96% of the suspended particles were eliminated by the single-stage UF membrane, which is slightly more than what previous studies for >20 kDa MWCO UF membranes for this task reported.[183,184] The percentage of COD rejection for both the single-stage and dual-stage treatments ranged from 90% to 97%. Comparable to the data for this study, literature values of 45%–69% COD rejection have been reported for low-strength public shower greywater for UF MWCO values between 30 and 400 kDa.[185] It was discovered that the 5 kDa UF membrane's COD rejection and membrane flux were more sensitive to temperature changes in the 25 °C–45 °C range than they were to pressure and crossflow rate changes in the 2.5–7.5 bar and 1–5 L min^{-1} ranges, respectively. When the temperature was raised from 25 °C to 45 °C, the percentage of COD rejection decreased by 7% (from 91%).

Hydrophilic polyvinylpyrollidone (PVP) modified polyethersulfone (PES) UF membranes were used to study laundry wastewater filtration.[39] Using a commercial PES membrane with a 10 kDa molecular weight in ultrafiltration, the PES/PVP membranes' capabilities were evaluated. For the modified PES membrane with high PVP concentration, a greater permeate flux of 55.2 L m^{-2} h^{-1} was obtained at 750 rpm of stirring speed and a TMP of 500 kPa. The findings demonstrated that, in comparison to other membranes, the PES membrane containing 10% PVP had a greater permeate flow, a better flux recovery, and less fouling. Because of the increased surface properties of the modified membranes, higher COD and TDS rejection of 88% and 82% were also noted. This demonstrated that surfactant, detergent, and oil from laundry effluent may be treated using modified PES membranes.

Based on Ciabattia *et al.* (2009) study, a prototype plant that was built and fed with 15 m^3 h^{-1} of untreated wastewater was used in a pre-industrial scale experiment to demonstrate a system for the purification and reuse of wastewater from an industrial laundry. Granular activated carbon (GAC) filtration is the post-treatment used on flat membranes for producing UF.[180] It was shown that, even with regard to the residual concentration of total surfactants (<2 mg L^{-1}), the GAC filter's outlet satisfies the regulatory requirements for discharge to surface waters in Italy when treatment parameters are adapted. Whiteness index data showed that the ultrafiltration post-treatment on flat membranes considerably reduced the residual pollutant content, permitting the treated effluents to be reused in certain home textile washing operations. Consequently, the whole system achieved a mean COD removal efficiency of 87%, and the TSS removal efficiency of the purification system was equally excellent at 98%.

4.9 Conclusion

Water scarcity has drawn attention to the reality that a sizeable portion of the global population lacks access to sanitary facilities and clean drinking water. In order to overcome these issues, effective treatment of industrial effluents and groundwater has recently become achievable because of recent advancements in membrane separation technology. To treat LWW, the UF membrane has become a promising option. Studies have evidenced the potential of UF for LWW treatment to produce treated water with a quality that is suitable for non-potable reuse. The major hurdle of UF for industrial applications is membrane fouling that has the potential to cause a permanent reduction in membrane permeability, in contrast to concentration polarization which exhibits the characteristic of being reversible. Currently, numerous proposed alterations have been put forth in order to enhance the efficacy of UF membranes in mitigating biofouling and fouling, as well as to optimize permeate flux and rejection. Despite the obstacles, continuing research holds hope for new discoveries. It is recommended that future research in this field should be focused on the development of robust membranes and the establishment of a more energy-efficient system for the practical application of UF for LWW treatment.

Acknowledgements

This work was supported by Ministry of Higher Education under Fundamental Research Grant Scheme (FRGS/1/2022/TK09/UTM/02/13)

References

1. UW Conference, *Historic UN Conference Marks Watershed Moment to Tackle Global Water Crisis and Ensure Water-Secure Future*, 2023.
2. D. Bonazzi, *Imminent Risk of a Global Water Crisis, Warns the UN World Water Development Report 2023*, 2023, 1–6.
3. UNESCO, The United Nations world water development report, 2016: Water and jobs: Executive Summary; 2016, 2016, available from: http://unesdoc.unesco.org/images/0024/002440/244040e.pdf
4. M. Yadav, R. Gupta and R. K. Sharma, Green and Sustainable Pathways for Wastewater Purification, in *Advances in Water Purification Techniques: Meeting the Needs of Developed and Developing Countries*, Elsevier Inc., 2019, pp. 355–383, DOI: 10.1016/B978-0-12-814790-0.00014-4.
5. N. Sims and B. Kasprzyk-Hordern, Future perspectives of wastewater-based epidemiology: Monitoring infectious disease spread and resistance to the community level, *Environ. Int.*, 2020, **139**(March), 105689.
6. R. A. Meyers, *Encyclopedia of Sustainability Science and Technology*, 2016.
7. Brenntag, Water Treatment Chemicals and Additives, 2023, available from: https://www.brenntag.com/en-ca/industries/water-treatment/#content_row_15.

8. HomeWater, What is Water Reclamation? Reuse, Recycle, Repurpose H_2O, 2022, available from: https://www.homewater.com/blog/what-is-water-reclamation-reuse-recycle-repurpose-h2o.
9. M. Patel and K. N. Sheth, A Study on Characterization & Treatment of Laundry Effluent, *IJIRST-Int. J. Innov. Res. Sci. Technol.*, 2017, **4**(1), 50–55.
10. M. A. Iribarnegaray, M. S. Rodriguez-Alvarez, L. B. Moraña, W. A. Tejerina and L. Seghezzo, Management challenges for a more decentralized treatment and reuse of domestic wastewater in metropolitan areas, *J. Water, Sanit. Hyg. Dev.*, 2018, **8**(1), 113–122.
11. S. A. Mousavi and F. Khodadoost, Effects of detergents on natural ecosystems and wastewater treatment processes: a review, *Environ. Sci. Pollut. Res.*, 2019, **26**(26), 26439–26448.
12. Z. M. Yaseen, T. T. Zigale, D. R. K. Tiyasha, S. Q. Salih and S. Awasthi, et al., Laundry wastewater treatment using a combination of sand filter, bio-char and teff straw media, *Sci. Rep.*, 2019, **9**(1), 1–11.
13. K. C. Ho, Y. H. Teow, J. Y. Sum, Z. J. Ng and A. W. Mohammad, Water pathways through the ages: Integrated laundry wastewater treatment for pollution prevention, *Sci. Total Environ.*, 2021, **760**, 143966.
14. J. Luo, Q. Zhang, L. Wu, Q. Feng, F. Fang and Z. Xue, et al., Promoting the anaerobic production of short-chain fatty acids from food wastes driven by the reuse of linear alkylbenzene sulphonates-enriched laundry wastewater, *Bioresour. Technol.*, 2019, **282**(February), 301–309.
15. V. V. Patil, P. R. Gogate, A. P. Bhat and P. K. Ghosh, Treatment of laundry wastewater containing residual surfactants using combined approaches based on ozone, catalyst and cavitation, *Sep. Purif. Technol.*, 2020, **239**(January), 116594.
16. M. Oteng-Peprah, M. A. Acheampong and N. K. deVries, Greywater Characteristics, Treatment Systems, Reuse Strategies and User Perception—a Review, *Water, Air, Soil Pollut.*, 2018, **229**(8), 255.
17. F. H. Al Hamedi, K. Kandhan, Y. Liu, M. Ren, A. Jaleel and M. A. M. Alyafei, Wastewater Irrigation: A Promising Way for Future Sustainable Agriculture and Food Security in the United Arab Emirates, *Water*, 2023, **15**(12), 1–18.
18. M. C. Collivignarelli, M. Carnevale Miino, M. Baldi, S. Manzi, A. Abbà and G. Bertanza, Removal of non-ionic and anionic surfactants from real laundry wastewater by means of a full-scale treatment system, *Process Saf. Environ. Prot.*, 2019, **132**, 105–115.
19. K. C. Ho, Y. H. Teow, J. Y. Sum, Z. J. Ng and A. W. Mohammad, Water pathways through the ages: Integrated laundry wastewater treatment for pollution prevention, *Sci. Total Environ.*, 2021, **760**, 143966.
20. M. Giagnorio, A. Amelio, H. Grüttner and A. Tiraferri, Environmental impacts of detergents and benefits of their recovery in the laundering industry, *J. Cleaner Prod.*, 2017, **154**, 593–601.
21. S. Mozia, M. Janus, P. Brożek, S. Bering, K. Tarnowski and J. Mazur, *et al.*, A system coupling hybrid biological method with UV/O_3 oxidation and membrane separation for treatment and reuse of industrial laundry wastewater, *Environ. Sci. Pollut. Res.*, 2016, **23**(19), 19145–19155.

22. F. Gholami, S. Zinadini, A. A. Zinatizadeh and A. R. Abbasi, TMU-5 metal-organic frameworks (MOFs) as a novel nanofiller for flux increment and fouling mitigation in PES ultrafiltration membrane, *Sep. Purif. Technol.*, 2018, **194**, 272–280.
23. C. Liu, H. Mao, J. Zheng and S. Zhang, Tight ultrafiltration membrane: Preparation and characterization of thermally resistant carboxylated cardo poly (arylene ether ketone)s (PAEK-COOH) tight ultrafiltration membrane for dye removal, *J. Membr. Sci.*, 2017, **530**, 1–10.
24. B. Pulido, S. Chisca and S. P. Nunes, Solvent and thermal resistant ultrafiltration membranes from alkyne-functionalized high-performance polymers, *J. Membr. Sci.*, 2018, **564**(July), 361–371.
25. M. Zahid, A. Rashid, S. Akram, Z. A. Rehan and W. Razzaq, A Comprehensive Review on Polymeric Nano-Composite Membranes for Water Treatment, *J. Membr. Sci. Technol.*, 2018, **08**(1), 1000179.
26. S. Imdad and R. K. Dohare, Preparation of ultrafiltration membrane from discarded polyethylene terephthalate bottles, *Environ. Sci. Pollut. Res.*, 2022, 42728–42737.
27. B. A. Pulido, O. S. Habboub, S. L. Aristizabal, G. Szekely and S. P. Nunes, Recycled Poly(ethylene terephthalate) for High Temperature Solvent Resistant Membranes, *ACS Appl. Polym. Mater.*, 2019, **1**(9), 2379–2387.
28. S. Rajesh and Z. V. P. Murthy, Ultrafiltration membranes from waste polyethylene terephthalate and additives: Synthesis and characterization, *Quim. Nova*, 2014, **37**(4), 653–657.
29. N. Trevor, We're Now At A Million Plastic Bottles Per Minute - 91% Of Which Are Not Recycled, *Forbes*, 2017, 5–9.
30. Z. Ding, L. Zhong, X. Wang and L. Zhang, Effect of lignin-cellulose nanofibrils on the hydrophilicity and mechanical properties of polyethersulfone ultrafiltration membranes, *High Perform. Polym.*, 2016, **28**(10), 1192–1200.
31. R. P. de O. Santos, P. F. Rossi, L. A. Ramos and E. Frollini, Renewable resources and a recycled polymer as raw materials: Mats from electrospinning of lignocellulosic biomass and PET solutions, *Polymers*, 2018, **10**(5), 538.
32. I. N. Strain, Q. Wu, A. M. Pourrahimi, M. S. Hedenqvist, R. T. Olsson and R. L. Andersson, Electrospinning of recycled PET to generate tough mesomorphic fibre membranes for smoke filtration, *J. Mater. Chem. A*, 2015, **3**(4), 1632–1640.
33. N. E. Zander, M. Gillan and D. Sweetser, Recycled PET nanofibers for water filtration applications, *Materials*, 2016, **9**(4), 1–10.
34. S. L. Soto Espinoza, C. R. Arbeitman, M. C. Clochard and M. Grasselli, Functionalization of nanochannels by radio-induced grafting polymerization on PET track-etched membranes, *Radiat. Phys. Chem.*, 2014, **94**(1), 72–75.
35. J. Hao, C. Xiao, T. Zhang, J. Zhao, Z. Fan and L. Chen, Preparation and Performance of PET-Braid-Reinforced Poly(vinylidene fluoride)/

Graphene Hollow-Fiber Membranes, *Ind. Eng. Chem. Res.*, 2016, **55**(7), 2174–2182.

36. Z. Zhou, L. F. Fang, S. Y. Wang and H. Matsuyama, Improving bonding strength between a hydrophilic coating layer and poly(ethylene terephthalate) braid for preparing mechanically stable braid-reinforced hollow fiber membranes, *J. Appl. Polym. Sci.*, 2018, **135**(14), 1–8.
37. P. Wang, M. Wang, F. Liu, S. Ding, X. Wang, G. Du, J. Liu, P. Apel, P. Kluth, C. Trautmann and Y. Wang, Ultrafast ion sieving using nanoporous polymeric membranes, *Nat. Commun.*, 2018, **9**(1), 569.
38. M. Manouchehri and A. Kargari, Water recovery from laundry wastewater by the cross flow microfiltration process: A strategy for water recycling in residential buildings, *J. Cleaner Prod.*, 2017, **168**, 227–238.
39. A. Sumisha, G. Arthanareeswaran, Y. Lukka Thuyavan, A. F. Ismail and S. Chakraborty, Treatment of laundry wastewater using polyethersulfone/polyvinylpyrollidone ultrafiltration membranes, *Ecotoxicol. Environ. Saf.*, 2015, **121**(2004), 174–179.
40. R. Mohamed, A. Kassim, M. Anda and S. Dallas, The effects of elements mass balance from turf grass irrigated with laundry and bathtub greywater, *Int. J. Appl. Environ. Sci.*, 2014, **9**(4), 2033–2049.
41. R. M. S. R. Mohamed, A. A. Wurochekke, C. M. Chan and A. H. bin M. Kassim, The Use of Natural Filter Media Added with Peat Soil for Household Greywater Treatment, *Int. J. Eng. Technol.*, 2014, **2**(3), 14–20.
42. T. Ramcharan and A. Bissessur, Analysis of Linear Alkylbenzene Sulfonate in Laundry Wastewater by HPLC-UV and UV-Vis Spectrophotometry, *J. Surfactants Deterg.*, 2016, **19**(1), 209–218.
43. J. Klaus, D. Edler, R. Beise-Zee, J. Blazejczak, R. Haum and M. Jänicke, *et al.*, *Lead Markets for Environmental Innovations*, 2005.
44. R. M. Mohamed, A. A. Al-Gheethi, J. Noramira, C. M. Chan, M. K. A. Hashim and M. Sabariah, Effect of detergents from laundry greywater on soil properties: a preliminary study, *Appl. Water Sci.*, 2018, **8**(1), 1–7.
45. N. L. Watiniasih, I. G. H. Purnama, G. Padmanabha, I. M. Merdana and I. N. G. Antara, Managing laundry wastewater, *IOP Conf. Ser.: Earth Environ. Sci.*, 2019, **248**(1), 8–13.
46. K. M. Khleifat, Biodegradation of sodium lauryl ether sulfate (SLES) by two different bacterial consortia, *Curr. Microbiol.*, 2006, **53**(5), 444–448.
47. S. Rejeki and M. A. R. Desrina, *Chronic Effects of Detergent Surfactant (Linear Alkylbenzene Sulfonate I Las) on the Growth and Survival Rate of Sea Bass (Lates Calcalifor Bloch) Larvae*, 2005.
48. D. S. Rajan, An evaluation of the effect of a detergent on dissolved oxygen consumption rate of Anabas testudineus, *Int. J. Fish. Aquat. Stud.*, 2015, **2**(6), 46–48.
49. M. P. Srivastava, R. Tiwari, N. Sharma and P. P. Division, Detergent and Sewage Phosphates entering into Lake Ecosystem and Its Impact on Aquatic Environment, *Int. J. Adv. Res.*, 2013, **1**(8), 1–6.

50. B. Jha and D. N. Singh, *Fly Ash Zeolites*, Springer, Singapore, 2016, available from: http://www.springer.com/in/book/9789811014024.
51. P. Pandey and B. Gopal, Effect of Detergents on the Growth of Two Aquatic Plants: Azolla pinnata and Hydrilla verticillata, *Environ. We Int. J. Sci. Technol.*, 2010, **5**, 107–114.
52. K. Kirshanov, R. Toms, G. Aliev, A. Naumova, P. Melnikov and A. Gervald, Recent Developments and Perspectives of Recycled Poly(ethylene terephthalate)-Based Membranes: A Review, *Membranes*, 2022, **12**(11), 1105.
53. T. K. G. Mohr, Environmental Investigation and Remediation: 1,4-Dioxane and other Solvent Stabilizers, in *Environmental Investigation and Remediation*, 2016.
54. M. S. Ayilara, O. S. Olanrewaju, O. O. Babalola and O. Odeyemi, Waste management through composting: Challenges and potentials, *Sustainability*, 2020, **12**(11), 1–23.
55. O. A. Alabi, K. I. Ologbonjaye, O. Awosolu and O. E. Alalade, Public and Environmental Health Effects of Plastic Wastes Disposal: A Review, *J. Toxicol. Risk Assess.*, 2019, **5**(2), 021.
56. U.S. Environmental Protection Agency (EPA), Risk Management for 1,4-Dioxane, Toxic Subst Control Act, 2020, 1–5.
57. F. Furtado, Earth Matters: Cleanup slow for Long Island's superfund sites, Isl Now, 2019, https://islandnow.net/earth-matters-cleanup-slow-for-long-islands-superfund-sites/.
58. E. P. Melián, D. E. Santiago, E. León, J. V. Reboso and J. A. Herrera-Melián, Treatment of laundry wastewater by different processes: Optimization and life cycle assessment, *J. Environ. Chem. Eng.*, 2023, **11**(2), 109302.
59. T. H. Kim, C. Park and S. Kim, Water recycling from desalination and purification process of reactive dye manufacturing industry by combined membrane filtration, *J. Cleaner Prod.*, 2005, **13**(8), 779–786.
60. B. Pirzadeh, *Physical Wastewater Treatment*, IntechOpen, 2022.
61. D. Christova-Boal, R. E. Eden and S. McFarlane, An investigation into greywater reuse for urban residential properties, *Desalination*, 1996, **106**(1–3), 391–397.
62. O. Lade and Z. Gbagba, Sustainable water supply: Potential of recycling laundry wastewater for domestic use, *J. Civ. Eng. Environ. Sci.*, 2017, **4**, 56–60.
63. Y. Liu, X. Wang, F. Yang and X. Yang, Excellent antimicrobial properties of mesoporous anatase TiO_2 and Ag/TiO_2 composite films, *Microporous Mesoporous Mater.*, 2008 Sep 1, **114**(1–3), 431–439.
64. N. Schouten, L. G. J. van der Ham, G. J. W. Euverink and A. B. de Haan, Selection and evaluation of adsorbents for the removal of anionic surfactants from laundry rinsing water, *Water Res.*, 2007, **41**(18), 4233–4241.
65. C. Zhang, H. Wen, Y. Huang and W. Shi, Adsorption of anionic surfactants from aqueous solution by high content of primary amino crosslinked chitosan microspheres, *Int. J. Biol. Macromol.*, 2017, **97**, 635–641.

66. J. N. Hakizimana, B. Gourich, M. Chafi, Y. Stiriba, C. Vial and P. Drogui, *et al.*, Electrocoagulation process in water treatment: A review of electrocoagulation modeling approaches, *Desalination*, 2017, **404**, 1–21.
67. F. Janpoor, A. Torabian and V. Khatibikamal, Treatment of laundry waste-water by electrocoagulation, *J. Chem. Technol. Biotechnol.*, 2011, **86**(8), 1113–1120.
68. K. W. Jung, M. J. Hwang, D. S. Park and K. H. Ahn, Performance evaluation and optimization of a fluidized three-dimensional electrode reactor combining pre-exposed granular activated carbon as a moving particle electrode for greywater treatment, *Sep. Purif. Technol.*, 2015, **156**, 414–423.
69. A. M. Abdel-Kader, Studying the efficiency of grey water treatment by using rotating biological contactors system, *J. King Saud Univ., Eng. Sci.*, 2013, **25**(2), 89–95.
70. E. R. Baumann, W. D. Hopping and F. D. Warner, Field evaluation of the treatability of type A zeolite in a trickling filter plant, *Water Res.*, 1981, **15**(7), 889–901.
71. N. Shamabadi, H. Bakhtiari, N. Kochakian and M. Farahani, The Investigation and Designing of an Onsite Grey Water Treatment Systems at Hazrat-e-Masoumeh University, Qom, IRAN, *Energy Procedia*, 2015, **74**, 1337–1346.
72. R. Tombola, G. Buttiglieri, M. Auset and R. Gonzalez-Olmos, Recycled corrugated wire hose cover as biological carriers for greywater treatment in a sequential batch biofilm reactor, *J. Environ. Manage.*, 2019, **240**(July 2018), 475–484.
73. M. Khalil and Y. Liu, Greywater biodegradability and biological treatment technologies: A critical review, *Int. Biodeterior. Biodegrad.*, 2021, **161**(December 2020), 105211.
74. M. F. Carosia, D. Y. Okada, I. K. Sakamoto, E. L. Silva and M. B. A. Varesche, Microbial characterization and degradation of linear alkylbenzene sulfonate in an anaerobic reactor treating wastewater containing soap powder, *Bioresour. Technol.*, 2014, **167**, 316–323.
75. V. B. Centurion, A. G. L. Moura, T. P. Delforno, D. Y. Okada, V. P. Dos Santos and M. B. A. Varesche, *et al.*, Anaerobic co-digestion of commercial laundry wastewater and domestic sewage in a pilot-scale EGSB reactor: The influence of surfactant concentration on microbial diversity, *Int. Biodeterior. Biodegrad.*, 2018, **127**(September 2017), 77–86.
76. I. C. S. Duarte, L. L. Oliveira, M. S. Mayor, D. Y. Okada and M. B. A. Varesche, Degradation of detergent (linear alkylbenzene sulfonate) in an anaerobic stirred sequencing-batch reactor containing granular biomass, *Int. Biodeterior. Biodegrad.*, 2010, **64**(2), 129–134.
77. T. P. Delforno, A. G. L. Moura, D. Y. Okada, I. K. Sakamoto and M. B. A. Varesche, Microbial diversity and the implications of sulfide levels in an anaerobic reactor used to remove an anionic surfactant from laundry wastewater, *Bioresour. Technol.*, 2015, **192**, 37–45.

78. T. P. Delforno, A. G. L. Moura, D. Y. Okada and M. B. A. Varesche, Effect of biomass adaptation to the degradation of anionic surfactants in laundry wastewater using EGSB reactors, *Bioresour. Technol.*, 2014, **154**, 114–121.
79. A. B. Esteban García, K. Szymański, S. Mozia and J. A. Sánchez Pérez, Treatment of laundry wastewater by solar photo-Fenton process at pilot plant scale, *Environ. Sci. Pollut. Res.*, 2021, **28**(7), 8576–8584.
80. J. A. Garrido-Cardenas, B. Esteban-García, A. Agüera, J. A. Sánchez-Pérez and F. Manzano-Agugliaro, Wastewater treatment by advanced oxidation process and their worldwide research trends, *Int. J. Environ. Res. Public Health*, 2020, **17**(1), 170.
81. P. Gautam, S. Kumar and S. Lokhandwala, Advanced oxidation processes for treatment of leachate from hazardous waste landfill: A critical review, *J. Cleaner Prod.*, 2019, **237**, 117639.
82. C. G. Joseph, Y. H. Taufiq-Yap, N. A. Affandi, J. L. H. Nga and V. Vijayan, Photocatalytic treatment of detergent-contaminated wastewater: A short review on current progress. Korean, *J. Chem. Eng.*, 2022, **39**(3), 484–498.
83. M. Von Sperling, *Principles of Biological Wastewater Treatment*, UFMG, Belo Horizonte/MG, 2005, vol. 3.
84. D. B. Miklos, C. Remy, M. Jekel, K. G. Linden, J. E. Drewes and U. Hübner, Evaluation of advanced oxidation processes for water and wastewater treatment – A critical review, *Water Res.*, 2018, **139**, 118–131.
85. M. A. Oturan and J. J. Aaron, Advanced oxidation processes in water/wastewater treatment: Principles and applications. A review, *Crit. Rev. Environ. Sci. Technol.*, 2014, **44**(23), 2577–2641.
86. E. M. Cuerda-Correa, M. F. Alexandre-Franco and C. Fernández-González, Advanced oxidation processes for the removal of antibiotics from water, An overview, *Water*, 2020, **12**(1), 102.
87. J. Wang and Z. Bai, Fe-based catalysts for heterogeneous catalytic ozonation of emerging contaminants in water and wastewater, *Chem. Eng. J.*, 2017, **312**, 79–98.
88. X. Huang, P. Gui and Y. Qian, Domestic wastewater treatment using a submerged membrane bioreactor, in *Progress in Biotechnology*, Elsevier Masson SAS, 2000, vol. 16, pp. 163–168, DOI: 10.1016/S0921-0423(00)80030-1.
89. E. Mahmoudi, L. Y. Ng, W. L. Ang, Y. H. Teow and A. W. Mohammad, Improving membrane bioreactor performance through the synergistic effect of silver-decorated graphene oxide in composite membranes, *J. Water Process Eng.*, 2020, **34**(January), 101169.
90. M. Y. Ashfaq and H. Qiblawey, Laundry wastewater treatment using ultrafiltration under different operating conditions, *AIP Conf. Proc.*, 2018, **2018**, 2022.
91. E. Siswoyo, A. W. Utari and L. G. N. Mungkari, Adsorption Combined Phytoremediation System for Treatment of Laundry Wastewater, *MATEC Web Conf.*, 2019, **280**, 05002.

92. J. Luo, L. Ding, Y. Wan and M. Y. Jaffrin, Flux decline control in nanofiltration of detergent wastewater by a shear-enhanced filtration system, *Chem. Eng. J.*, 2012, **181–182**, 397–406.
93. J. Guilbaud, A. Massé, Y. Andrès, F. Combe and P. Jaouen, Influence of operating conditions on direct nanofiltration of greywaters: Application to laundry water recycling aboard ships, *Resour., Conserv. Recycl.*, 2012, **62**, 64–70.
94. X. Zhang, D. Taylor, R. Thomas, J. Penfold and I. Tucker, Modifying the adsorption properties of anionic surfactants onto hydrophilic silica using the pH dependence of the polyelectrolytes PEI, ethoxylated PEI, and polyamines, *Langmuir*, 2011, **27**(7), 3569–3577.
95. C. C. Borghi, M. Fabbri, M. Fiorini, M. Mancini and P. L. Ribani, Magnetic removal of surfactants from wastewater using micrometric iron oxide powders, *Sep. Purif. Technol.*, 2011, **83**(1), 180–188.
96. E. Yüksel, I. A. Şengil and M. Özacar, The removal of sodium dodecyl sulfate in synthetic wastewater by peroxi-electrocoagulation method, *Chem. Eng. J.*, 2009, **152**(2–3), 347–353.
97. P. N. Rahardjo, Teknologi Pengelolaan Limbah Cair Yang Ideal, *JAI*, 2006, **2**(1), 66–71.
98. B. Yang, G. Chen and G. Chen, Submerged membrane bioreactor in treatment of simulated restaurant wastewater, *Sep. Purif. Technol.*, 2012, **88**, 184–190.
99. J. Dilafroza, K. P. Ashok and N. K. Azra, Efficiency evaluation of three fluidised aerobic bioreactor based sewage treatment plants in Kashmir Valley, *Afr. J. Biotechnol.*, 2013, **12**(17), 2224–2233.
100. H. Chan, High performance achieved by microbes to separate laundry effluents resulting in producing high water quality in a compact area, *Sep. Purif. Technol.*, 2012, **90**, 101–108.
101. C. Corada-Fernández, E. González-Mazo and P. A. Lara-Martín, Evaluation of the anaerobic biodegradation of linear alkylbenzene sulfonates (LAS) using OECD 308 water/sediment systems, *J. Hazard. Mater.*, 2018, **360**(March), 24–31.
102. L. L. de Oliveira, R. B. Costa, D. Y. Okada, D. V. Vich, I. C. S. Duarte and E. L. Silva, *et al.*, Anaerobic degradation of linear alkylbenzene sulfonate (LAS) in fluidized bed reactor by microbial consortia in different support materials, *Bioresour. Technol.*, 2010, **101**(14), 5112–5122.
103. U. Merrettig-Bruns and E. Jelen, Anaerobic biodegradation of detergent surfactants, *Materials*, 2009, **2**(1), 181–206.
104. J. Ji, S. Sakuma, J. Ni, Y. Chen, Y. Hu and A. Ohtsu, et al., Application of two anaerobic membrane bioreactors with different pore size membranes for municipal wastewater treatment, *Sci. Total Environ.*, 2020, **745**, 140903.
105. F. Mega, Pengolahan Limbah Cair Domestik Dengan Biofilter Aerob Menggunakan Media Bioball Dan Tanaman Kiambang, *J. Teknol. Lingkung Lahan Basah*, 2013, **1**(1), 1–10.
106. W. P. Utomo, Z. V. Nugraheni, A. Rosyidah, O. M. Shafwah, L. K. Naashihah and N. Nurfitria, *et al.*, Penurunan Kadar Surfaktan Anionik dan

Fosfat dalam Air Limbah Laundry di Kawasan Keputih, Surabaya menggunakan Karbon Aktif, *Akta Kim. Indones*, 2018, **3**(1), 127.

107. A. K. Choudhary and S. Kumar, Constructed Wetlands for Wastewater Treatment, in *Manag Wastewater Sludge New Approaches*, 2011.
108. J. J. Pignatello, E. Oliveros and A. MacKay, Advanced oxidation processes for organic contaminant destruction based on the fenton reaction and related chemistry, *Crit. Rev. Environ. Sci. Technol.*, 2006, **36**(1), 1–84.
109. R. Ameta, A. K. Chohadia, A. Jain and P. B. Punjabi, Fenton and Photo-Fenton Processes, in *Advanced Oxidation Processes for Wastewater Treatment, Emerging Green Chemical Technology*, 2018, pp. 49–87.
110. C. Amor, L. Marchão, M. S. Lucas and J. A. Peres, Application of advanced oxidation processes for the treatment of recalcitrant agro-industrial wastewater: A review, *Water*, 2019, **11**(2), 205.
111. D. B. Miklos, R. Hartl, P. Michel, K. G. Linden, J. E. Drewes and U. Hübner, UV/H_2O_2 process stability and pilot-scale validation for trace organic chemical removal from wastewater treatment plant effluents, *Water Res.*, 2018, **136**, 169–179.
112. M. Klavarioti, D. Mantzavinos and D. Kassinos, Removal of residual pharmaceuticals from aqueous systems by advanced oxidation processes, *Environ. Int.*, 2009, **35**(2), 402–417.
113. M. Umar, H. A. Aziz and M. S. Yusoff, Trends in the use of Fenton, electro-Fenton and photo-Fenton for the treatment of landfill leachate, *Waste Manag.*, 2010, **30**(11), 2113–2121.
114. Y. Yu, Z. Chen, Z. Guo, Z. Liao, L. Yang and J. Wang, *et al.*, Removal of refractory contaminants in municipal landfill leachate by hydrogen, oxygen and palladium: A novel approach of hydroxyl radical production, *J. Hazard. Mater.*, 2015, **287**, 349–355.
115. H. Hansson, F. Kaczala, M. Marques and W. Hogland, Photo-Fenton and Fenton oxidation of recalcitrant industrial wastewater using nanoscale zero-valent iron, *Int. J. Photoenergy*, 2012, **2012**, 531076.
116. S. Miralles-Cuevas, I. Oller, A. Agüera, J. A. Sánchez Pérez and S. Malato, Strategies for reducing cost by using solar photo-Fenton treatment combined with nanofiltration to remove microcontaminants in real municipal effluents: Toxicity and economic assessment, *Chem. Eng. J.*, 2017, **318**, 161–170.
117. J. Chen, M. Liu, J. Zhang, X. Ying and L. Jin, Photocatalytic degradation of organic wastes by electrochemically assisted TiO_2 photocatalytic system, *J. Environ. Manage.*, 2004, **70**(1), 43–47.
118. M. R. Bilad, N. I. Mat Nawi, D. D. Subramaniam, N. Shamsuddin, A. L. Khan and J. Jaafar, *et al.*, Low-pressure submerged membrane filtration for potential reuse of detergent and water from laundry wastewater, *J. Water Process Eng.*, 2020, **36**(April), 101264.
119. F. Buchheister, J. Hoinkis, S. Muth and V. Panten, LIWATEC - laundry innovative waste water recycling technology, *Desalination*, 2006, **199**(1–3), 76–77.

120. C. Nicolaidis and I. Vyrides, Closing the water cycle for industrial laundries: An operational performance and techno-economic evaluation of a full-scale membrane bioreactor system, *Resour., Conserv. Recycl.*, 2014, **92**, 128–135.
121. K. Oktor and D. Çelik, Treatment of wash basin and bathroom greywater with Chlorella variabilis and reusability, *J. Water Process Eng.*, 2019, **31**(December 2018), 100857.
122. W. C. Chong, A. W. Mohammad, E. Mahmoudi, Y. T. Chung, K. F. Kamarudin and M. S. Takriff, Nanohybrid membrane in algal-membrane photoreactor: Microalgae cultivation and wastewater polishing, *Chin. J. Chem. Eng.*, 2019, **27**(11), 2799–2806.
123. Z. M. Á López, L. B. Suárez Pérez, L. Reynoso-Cuevas and N. Funamizu, Pre-filtration for enhancing direct membrane filtration of graywater from washing machine discharges, *Ecol. Eng.*, 2014, **64**, 116–119.
124. M. C. Von, *Polymers in Information Storage Technology*, Technomic Publ Company, Inc., 1987, vol. 38(12), p. 691.
125. J. M. Laîné, D. Vial and P. Moulart, Status after 10 years of operation - overview of UF technology today, *Desalination*, 2000, **131**(1–3), 17–25.
126. J. S. S. Yadav, S. Yan, C. M. Ajila, J. Bezawada, R. D. Tyagi and R. Y. Surampalli, Food-grade single-cell protein production, characterization and ultrafiltration recovery of residual fermented whey proteins from whey, *Food Bioprod. Process.*, 2016, **99**, 156–165.
127. Z. Zhu, F. Yuan, Z. Xu, W. Wang, X. Di and F. J. Barba, *et al.*, Stirring-assisted dead-end ultrafiltration for protein and polyphenol recovery from purple sweet potato juices: Filtration behavior investigation and HPLC-DAD-ESI-MS2 profiling, *Sep. Purif. Technol.*, 2016, **169**, 25–32.
128. C. M. Galanakis, Separation of functional macromolecules and micromolecules: From ultrafiltration to the border of nanofiltration, *Trends Food Sci. Technol.*, 2015, **42**(1), 44–63.
129. X. Shi, G. Tal, N. P. Hankins and V. Gitis, Fouling and cleaning of ultrafiltration membranes: A review, *J. Water Process Eng.*, 2014, **1**, 121–138.
130. S. Lee, M. Ihara, N. Yamashita and H. Tanaka, Improvement of virus removal by pilot-scale coagulation-ultrafiltration process for wastewater reclamation: Effect of optimization of pH in secondary effluent, *Water Res.*, 2017, **114**, 23–30.
131. W. Zou and K. R. Davey, An integrated two-step Fr 13 synthesis - demonstrated with membrane fouling in combined ultrafiltration-osmotic distillation (UF-OD) for concentrated juice, *Chem. Eng. Sci.*, 2016, **152**, 213–226.
132. X. Zhou, J. Liang, Y. Zhang, H. Zhao, Y. Guo and S. Shi, Separation and purification of α-glucosidase inhibitors from Polygonatum odoratum by stepwise high-speed counter-current chromatography combined with Sephadex LH-20 chromatography target-guided by ultrafiltration-HPLC screening, *J. Chromatogr. B: Anal. Technol. Biomed. Life Sci.*, 2015, **985**, 149–154.

133. K. M. Barry, T. G. Dinan and P. M. Kelly, Pilot scale production of a phospholipid-enriched dairy ingredient by means of an optimised integrated process employing enzymatic hydrolysis, ultrafiltration and super-critical fluid extraction, *Innovative Food Sci. Emerging Technol.*, 2017, **41**, 301–306.
134. A. W. Mohammad, C. Y. Ng, Y. P. Lim and G. H. Ng, Ultrafiltration in Food Processing Industry: Review on Application, Membrane Fouling, and Fouling Control, *Food Bioprocess Technol.*, 2012, **5**(4), 1143–1156.
135. R. Rohani, M. Hyland and D. Patterson, A refined one-filtration method for aqueous based nanofiltration and ultrafiltration membrane molecular weight cut-off determination using polyethylene glycols, *J. Membr. Sci.*, 2011, **382**(1–2), 278–290.
136. X. Li, L. Jiang and H. Li, Application of Ultrafiltration Technology in Water Treatment, *IOP Conf. Ser.: Earth Environ. Sci.*, 2018, **186**(3), 012009.
137. Q. Wu and Q. Chen, Application of membrane separation technology in water treatment process, *IOP Conf. Ser.: Earth Environ. Sci.*, 2020, **508**(1), 012048.
138. U. M. Aliyu, S. Rathilal and Y. M. Isa, Membrane desalination technologies in water treatment: A review, *Water Pract. Technol.*, 2018, **13**(4), 738–752.
139. M. Amirilargani, M. Sadrzadeh and T. Mohammadi, Synthesis and characterization of polyethersulfone membranes, *J. Polym. Res.*, 2010, **17**(3), 363–377.
140. B. Chakrabarty, A. K. Ghoshal and M. K. Purkait, Preparation, characterization and performance studies of polysulfone membranes using PVP as an additive, *J. Membr. Sci.*, 2008, **315**(1–2), 36–47.
141. D. Y. Zuo, Y. Y. Xu, W. L. Xu and H. T. Zou, The influence of PEG molecular weight on morphologies and properties of PVDf asymmetric membranes, *Chin. J. Polym. Sci.*, 2008, **26**(4), 405–414.
142. E. Saljoughi and T. Mohammadi, Cellulose acetate (CA)/polyvinylpyrrolidone (PVP) blend asymmetric membranes: Preparation, morphology and performance, *Desalination*, 2009, **249**(2), 850–854.
143. S. Yang and Z. Liu, Preparation and characterization of polyacrylonitrile ultrafiltration membranes, *J. Membr. Sci.*, 2003, **222**(1–2), 87–98.
144. M. Amirilargani, A. Sabetghadam and T. Mohammadi, Polyethersulfone/polyacrylonitrile blend ultrafiltration membranes with different molecular weight of polyethylene glycol: Preparation, morphology and antifouling properties, *Polym. Adv. Technol.*, 2012, **23**(3), 398–407.
145. E. Drioli and L. Giorno, *Membrane Operations: Innovative Separations and Transformations*, Wiley-VCH, 2009.
146. F. Liu, N. A. Hashim, Y. Liu, M. R. M. Abed and K. Li, Progress in the production and modification of PVDF membranes, *J. Membr. Sci.*, 2011, **375**(1–2), 1–27.
147. I. Pinnau and B. D. Freeman, Formation and modification of polymeric membranes: Overview, *ACS Symp. Ser.*, 1999, **744**, 1–22.

148. B. Wang and Z. Lai, Finger-like voids induced by viscous fingering during phase inversion of alumina/PES/NMP suspensions, *J. Membr. Sci.*, 2012, **405–406**, 275–283.
149. M. M. Pendergast and E. M. V. Hoek, A review of water treatment membrane nanotechnologies, *Energy Environ. Sci.*, 2011, **4**(6), 1946–1971.
150. J. Yang, D. W. Li, Y. K. Lin, X. L. Wang, F. Tian and Z. Wang, Formation of a bicontinuous structure membrane of polyvinylidene fluoride in diphenyl ketone diluent via thermally induced phase separation, *J. Appl. Polym. Sci.*, 2008, **110**(1), 341–347.
151. P. L. Hanks, *Post Coarsening Effects on Membrane Microstructure*, 2008.
152. S. Kim and C. Park, Fouling behavior and cleaning strategies of ceramic ultrafiltration membranes for the treatment and reuse of laundry wastewater, *J. Water Process Eng.*, 2022, **48**(December 2021), 102840.
153. N. M. D'Souza and A. J. Mawson, Membrane cleaning in the dairy industry: A review, *Crit. Rev. Food Sci. Nutr.*, 2005, **45**(2), 125–134.
154. K. Kimura, Y. Hane, Y. Watanabe, G. Amy and N. Ohkuma, Irreversible membrane fouling during ultrafiltration of surface water, *Water Res.*, 2004, **38**(14–15), 3431–3441.
155. S. C. Tu, V. Ravindran and M. Pirbazari, A pore diffusion transport model for forecasting the performance of membrane processes, *J. Membr. Sci.*, 2005, **265**(1–2), 29–50.
156. J. Luo, L. Ding, B. Qi, M. Y. Jaffrin and Y. Wan, A two-stage ultrafiltration and nanofiltration process for recycling dairy wastewater, *Bioresour. Technol.*, 2011, **102**(16), 7437–7442.
157. L. J. Zeman and A. L. Zydney, *Microfiltration and Ultrafiltration | Principles and Applications*, 2017, 642 p.
158. R. Semiat and D. Hasson, Seawater and Brackish-Water Desalination with Membrane Operations, *Membr. Technol.*, 2010, **4**, 147–168.
159. A. I. Schäfer and A. G. Fane, *Nanofiltration | Principles, Applications, and New Materials*, WILEY-VCH, 2021, vol. 6, 128 p.
160. S. Srisurichan, R. Jiraratananon and A. G. Fane, Mass transfer mechanisms and transport resistances in direct contact membrane distillation process, *J. Membr. Sci.*, 2006, **277**(1–2), 186–194.
161. F. Li, J. Meng, J. Ye, B. Yang, Q. Tian and C. Deng, Surface modification of PES ultrafiltration membrane by polydopamine coating and poly(ethylene glycol) grafting: Morphology, stability, and anti-fouling, *Desalination*, 2014, **344**, 422–430.
162. N. R. Choudhury, A. G. Kannan and N. K. Dutta, Novel nanocomposites and hybrids for lubricating coating applications, in *Tribology and Interface Engineering Series*, Elsevier Ltd, 2008, ch. 21, vol. 55, pp. 501–542, DOI: 10.1016/S1572-3364(08)55021-X.
163. S. Zare and A. Kargari, Membrane properties in membrane distillation, *Emerging Technologies for Sustainable Desalination Handbook*, Elsevier Inc., 2018, pp. 107–156, DOI: 10.1016/B978-0-12-815818-0.00004-7.
164. L. C. Capozzi, F. M. Mehmood, M. Giagnorio, A. Tiraferri, M. Cerruti and M. Sangermano, Ultrafiltration Membranes Functionalized with

Polydopamine with Enhanced Contaminant Removal by Adsorption, *Macromol. Mater. Eng.*, 2017, **302**(5), 1–10.
165. K. Y. Kim, E. Yang, M. Y. Lee, K. J. Chae, C. M. Kim and I. S. Kim, Polydopamine coating effects on ultrafiltration membrane to enhance power density and mitigate biofouling of ultrafiltration microbial fuel cells (UF-MFCs), *Water Res.*, 2014, **54**, 62–68.
166. Y. H. Ding, M. Floren and W. Tan, Mussel-inspired polydopamine for bio-surface functionalization, *Biosurface Biotribology*, 2016, **2**(4), 121–136.
167. B. Van der Bruggen, Chemical Modification of Polyethersulfone Nanofiltration Membranes: A Review, *J. Appl. Polym. Sci.*, 2009, **114**(1), 630–642.
168. N. A. Alenazi, M. A. Hussein, K. A. Alamry and A. M. Asiri, Modified polyether-sulfone membrane: A mini review, *Des. Monomers Polym.*, 2017, **20**(1), 532–546.
169. B. Díez and R. Rosal, A critical review of membrane modification techniques for fouling and biofouling control in pressure-driven membrane processes, *Nanotechnol. Environ. Eng.*, 2020, **5**(2), 15.
170. A. Anvari, A. A. Yancheshme, M. Tavakolmoghadam, F. Rekabdar and A. Safekordi, Effect of TiO_2 nanoparticles on antifouling and separation properties of PVDF/PAN blend membrane, *Desalin. Water Treat.*, 2019, **154**(May), 92–100.
171. L. Y. Ng, A. W. Mohammad, C. P. Leo and N. Hilal, Polymeric membranes incorporated with metal/metal oxide nanoparticles: A comprehensive review, *Desalination*, 2013, **308**, 15–33.
172. E. S. Awad, T. M. Sabirova, N. A. Tretyakova, Q. F. Alsalhy, A. Figoli and I. K. Salih, A Mini-Review of Enhancing Ultrafiltration Membranes (UF) for Wastewater Treatment : Performance and Stability, *ChemEngineering*, 2021, **5**(3), 34.
173. R. Kumar and A. F. Ismail, Fouling control on microfiltration/ultrafiltration membranes: Effects of morphology, hydrophilicity, and charge, *J. Appl. Polym. Sci.*, 2015, **132**(21), 42042.
174. X. J. Lee, P. L. Show, T. Katsuda, W. H. Chen and J. S. Chang, Surface grafting techniques on the improvement of membrane bioreactor: State-of-the-art advances, *Bioresour. Technol.*, 2018, **269**, 489–502.
175. N. De Geyter and R. Morent, Cold plasma surface modification of biodegradable polymer biomaterials, in *Biomaterials for Bone Regeneration: Novel Techniques and Applications*, Woodhead Publishing Limited, 2014, pp. 202–224, DOI: 10.1533/9780857098104.2.202.
176. L. Braeken, B. Van Der Bruggen and C. Vandecasteele, Regeneration of brewery waste water using nanofiltration, *Water Res.*, 2004, **38**(13), 3075–3082.
177. B. D. P. Luogo, T. Salim, W. Zhang, N. B. Hartmann, F. Malpei and V. M. Candelario, Reuse of Water in Laundry Applications with Micro-and Ultrafiltration Ceramic Membrane, *Membranes*, 2022, **12**(2), 1–11.

178. C. O. C. Nascimento, M. T. Veit, S. M. Palácio, G. C. Gonçalves and M. R. Fagundes-Klen, Combined Application of Coagulation/Flocculation/Sedimentation and Membrane Separation for the Treatment of Laundry Wastewater, *Int. J. Chem. Eng.*, 2019, **2019**, 324710.
179. M. Y. Ashfaq, T. Wang, H. Qiblawey, I. A. Reesh and S. Judd, Recycling of hospital laundry wastewater using membrane technology, *Desalin. Water Treat.*, 2017, **60**, 122–128.
180. I. Ciabattia, F. Cesaro, L. Faralli, E. Fatarella and F. Tognotti, Demonstration of a treatment system for purification and reuse of laundry wastewater, *Desalination*, 2009, **245**(1–3), 451–459.
181. S. Barredo-Damas, M. I. Alcaina-Miranda, M. I. Iborra-Clar, A. Bes-Piá, J. A. Mendoza-Roca and A. Iborra-Clar, Study of the UF process as pretreatment of NF membranes for textile wastewater reuse, *Desalination*, 2006, **200**(1–3), 745–747.
182. T. Wang, H. Qiblawey, S. Judd, A. Benamor, M. S. Nasser and A. Mohammadian, Fabrication of high flux nanofiltration membrane via hydrogen bonding based co-deposition of polydopamine with poly(vinyl alcohol), *J. Membr. Sci.*, 2018, **552**(December 2017), 222–233.
183. M. Ernst, A. Sperlich, X. Zheng, Y. Gan, J. Hu and X. Zhao, *et al.*, An integrated wastewater treatment and reuse concept for the Olympic Park 2008, Beijing, *Desalination*, 2007, **202**(1–3), 293–301.
184. J. Benner, D. E. Helbling, H. P. E. Kohler, J. Wittebol, E. Kaiser and C. Prasse, *et al.*, Is biological treatment a viable alternative for micropollutant removal in drinking water treatment processes?, *Water Res.*, 2013, **47**(16), 5955–5976.
185. G. Ramon, M. Green, R. Semiat and C. Dosoretz, Low strength graywater characterization and treatment by direct membrane filtration, *Desalination*, 2004, **170**(3), 241–250.

CHAPTER 5

TiO_2–Graphitic Carbon Nitride-based Nanocomposites for the Degradation of Emerging Pollutants

DEVAGI KANAKARAJU[a*] AND LIM YING CHIN[b]

[a]Faculty of Resource Science and Technology, Universiti Malaysia Sarawak, 94300 Kota Samarahan, Malaysia; [b]Faculty of Applied Sciences, Universiti Teknologi MARA, 40450 Shah Alam, Selangor, Malaysia
*E-mail: kdevagi@unimas.my

5.1 Introduction

In recent years, the proliferation of emerging pollutants in the environment has posed significant challenges to global ecosystems and human health. These pollutants, often stemming from industrial activities, agriculture, and urbanization, exhibit resistance to conventional treatment methods, necessitating the development of innovative strategies for their removal. Among these strategies, heterogeneous photocatalysis has emerged as a promising approach due to its effectiveness in degrading a wide range of organic contaminants under ambient conditions.

In the realm of photocatalytic materials, titanium dioxide (TiO_2) has garnered considerable attention owing to its excellent photocatalytic properties, chemical stability, and abundance. However, it possesses a large band gap that is inefficient in the exploitation of visible light and is limited to ultraviolet

Chemistry in the Environment Series No. 14
Advanced Materials for Emerging Water Pollutant Removal
Edited by Pei Sean Goh, Devagi Kanakaraju, Anwar Iqbal and Ahmad Fauzi Ismail

Published by the Royal Society of Chemistry, www.rsc.org

light. The inherent instability of TiO_2 often leads to the aggregation of TiO_2 nanoparticles and the formation of suspensions. The aggregation of active sites on the TiO_2 surface results in a hindrance to the absorption of photon energy, leading to a reduction in the photodegradation rate. Hence, surface modification of TiO_2 becomes necessary. Studies have demonstrated that the combination of TiO_2 with a carbonaceous material, graphitic carbon nitride (gC_3N_4), a metal-free conductor is highly preferred due to its narrow band gap (~2.7 eV), making it responsive to visible light. Furthermore, gC_3N_4 possesses high charge mobility which exhibits high photocatalytic activity. This is due to its unique electronic structure, which enables it to efficiently generate electron–hole pairs and facilitate redox reactions. By combining the unique properties of TiO_2 and gC_3N_4, synergistic effects can be achieved, leading to improved charge separation, an extended light absorption range, and enhanced photocatalytic performance for the degradation of various emerging pollutants.

This chapter highlights the modification of binary TiO_2–gC_3N_4 and TiO_2–gC_3N_4-based nanostructures *via* various synthesis methods and their applications for the removal of emerging pollutants.

5.2 Titanium Dioxide

TiO_2 is extensively applied in many areas such as cosmetics, medical research, self-cleaning technology, and water purification. The primary contributing advantages to its wide applications are strong oxidizing power, high chemical stability, its non-toxic nature, and low cost. The application of TiO_2 for the remediation of emerging pollutants such as pharmaceuticals and personal care products, endocrine-disrupting contaminants, pesticides, and dyes is known to be driven by the ability of the nanoparticles to completely oxidize or mineralize organic pollutants without producing toxic by-products. The production of reactive oxygen species (ROS) is important to initiate mineralization and transformation of the pollutants to water, carbon dioxide, inorganic ions, and heteroatoms.

5.3 Crystalline Structures of Titanium Dioxide

The three important crystalline phases of TiO_2 are anatase, brookite, and rutile.[1] Table 5.1 summarizes the properties of the crystalline phases of TiO_2. All three crystal phases have a TiO_6 octahedron structure as a result of the six O_2^- ions surrounding the titanium (Ti^{4+}) atoms. Both anatase TiO_2 and rutile TiO_2 have tetragonal crystal systems while brookite has an orthorhombic crystal system.[2] Structure differences may be ascribed to different photocatalytic activities of TiO_2 polymorphs.[3] Among these TiO_2 polymorphs, anatase and rutile are widely used as photocatalysts due to their high stabilities and good photoreactive properties.[2] The complicated phase of brookite with high unit cell volume and minimum density leads to its limited application. Pure brookite is not suitable for photodegradation or photosynthetic reaction due to its laborious preparation method.[4]

Table 5.1 Crystal structure and properties of three different phases of TiO_2.

Properties	Phases of TiO_2			References
	Anatase	Brookite	Rutile	
Crystal structure	Tetragonal	Orthorhombic	Tetragonal	Etacheri *et al.*[5]
Atoms per unit cells	6	8	6	Gupta & Tripathi[6]
Packing density (g cm^{-3})	3.79	3.99	4.13	Luo *et al.*[3]
Band gap (eV)	3.2	3.4	3.0	Etacheri *et al.*[5]

The anatase phase is the preferred photocatalyst because of its chemical properties namely higher electron mobility, low dielectric constant and lower density.[6,7] The superior photocatalytic activity of anatase TiO_2 could also be explained by comparing the optical and physical properties of the polymorphs. The band gap of anatase, brookite, and rutile are 3.2, 3.4, and 3.0 eV, respectively. This band gap shows that only the rutile phase could absorb radiation near visible light while others could only absorb UV radiation.[8] Although rutile can absorb visible light, the photocatalytic activity of anatase is relatively high compared to rutile due to its conduction band position which provides stronger reduction power than rutile.[8] In addition, the lower packing density of anatase TiO_2 (3.79 g cm^{-3}) compared to rutile (4.13 g cm^{-3}) and brookite (3.99 g cm^{-3}) implies the relatively better photoactivity of the former.[3] Anatase also possesses the lightest average effective mass of photoexcited electrons and holes compared to rutile and brookite which translates the fastest migration of photoexcited electrons and holes from the interior to the anatase surface resulting in lower recombination rates of electron and hole pairs.[9]

The resulting crystal phase of TiO_2 depends on the annealing temperature, synthesis method, and starting material.[10] With regard to the annealing temperature, rutile is the most stable phase at most temperatures[10] whilst anatase is the most stable phase at 0 K.[11] The metastable phases of anatase and brookite allow them to transform into the rutile phase upon heating.[10] The rate of transformation of anatase or brookite to the rutile phase increases exponentially with the increase in temperature.[12] Bakri *et al.*[10] reported that anatase was obtained at a temperature of 300 °C and transformed into the rutile phase at 900 °C. Both brookite and anatase exothermically transform to the rutile phase at a temperature exceeding 600 °C.[5]

5.4 Mechanism of TiO_2 Photocatalysis

As a photocatalyst, TiO_2 can induce a series of oxidative–reductive reactions as shown in eqn (5.1)–(5.12).[8,13] Figure 5.1 shows the TiO_2 photocatalysis reaction in the presence of a pollutant (PH). TiO_2 photocatalysis requires photon (hv) absorption from an excitation wavelength of 388 nm[14] to induce electron (e^-) excitation from the valence band (VB) to the conduction band (CB) and eventually form positive hole (h^+) in the valence band (eqn (5.1)).[7]

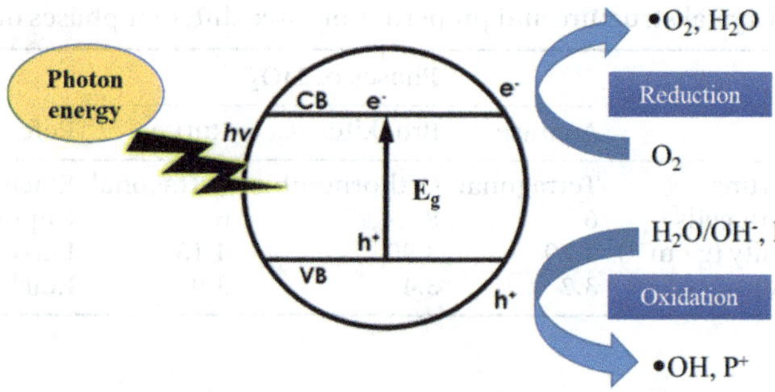

Figure 5.1 Mechanism of TiO_2 photocatalysis.

The valence band and the conduction band consist of overlapping oxygen 2p orbitals and 3d orbitals of Ti^{4+} cations, respectively.[15] The energy difference between the conduction band and the valence band is known as the band gap energy, E_g (Figure 5.1).

The electron–hole pairs would undergo various oxidative and reductive reactions with the species adsorbed on the TiO_2 surface.[7] Positive holes at the valence band oxidize the adsorbed water (H_2O) and hydroxyl ion (OH^-) to generate hydroxyl radicals ($^{\bullet}OH$) (eqn (5.2) and (5.3)).[7] The positive hole could also directly react and degrade the adsorbed pollutant (RH) but this reaction is very unlikely to occur (eqn (5.4)).[8] The excited electrons at the conduction band react with oxygen (O_2) generating superoxide anion radical ($^{\bullet}O_2^-$) (eqn (5.5)).[7] The $^{\bullet}O_2^-$ radicals can scavenge water and OH^- to generate $^{\bullet}OH$ radicals (eqn (5.6) and (5.7)). In some cases, the photoexcited electron and hole pairs might also undergo recombination without participating in the subsequent photocatalytic reactions and release absorbed light energy as heat.[8,16] Nevertheless, the recombination of electron–hole pairs could be avoided by the addition of scavenger or a surface defect to trap the electron and hole during the photocatalysis process[8]

Photoexcitation:

$$TiO_2 + h\nu \rightarrow TiO_2e^-(CB) + TiO_2h^+(VB) \quad (5.1)$$

Oxidation of water:

$$TiO_2(h^+) + H_2O_{ads} \rightarrow TiO_2 + \bullet OH_{ads} + H^+ \quad (5.2)$$

Oxidation of hydroxyl ions:

$$TiO_2(h^+) + OH^- \rightarrow TiO_2 + \bullet OH_{ads} \quad (5.3)$$

Direct photo holes:

$$PH + TiO_2(h^+) \rightarrow P^+ \rightarrow Intermediate(s) \rightarrow CO_2 + H_2O \quad (5.4)$$

Reduction of oxygen:

$$TiO_2(e^-) + (O_2) \rightarrow TiO_2 + \bullet O_2^- \quad (5.5)$$

Superoxide radical scavenge on water:

$$\bullet O_2^- + 2H_2O_{ads} \rightarrow 2\bullet OH_{ads} + 2OH^- \quad (5.6)$$

Superoxide radical scavenge on hydroxyl ions:

$$\bullet O_2^- + 2OH^- \rightarrow 2\bullet OH_{ads} + O_2 \quad (5.7)$$

The superoxide radical ($^{\bullet}O_2^-$) can be further protonated to form hydroperoxyl radical ($^{\bullet}OOH$) (eqn (5.8)).[13] The hydroperoxyl radicals can be reduced to form hydroperoxyl anion (OOH^-) and further oxidized in the presence of hydrogen ions (H^+) to form hydrogen peroxide (H_2O_2) (eqn (5.9) and (5.10)).[13] These H_2O_2 molecules can generate more $^{\bullet}OH$ radicals leading to a higher photodegradation rate (eqn (5.11)).[8] The generated $^{\bullet}OH$ and $^{\bullet}O_2^-$ radicals can rapidly attack the adsorbed pollutant (PH) on the TiO_2 surface[8] (eqn (5.12)). The photocatalytic processes can mineralize the pollutants by continuously breaking down pollutants until no residue of the original materials remains[8]

Protonation of superoxides:

$$\bullet O_2^- + \bullet OH_{ads} \rightarrow \bullet OOH \quad (5.8)$$

Co-scavenging of e^-:

$$\bullet OOH + e^- \rightarrow OOH^- \quad (5.9)$$

Formation of H_2O_2:

$$OOH^- + H^+ \rightarrow H_2O_2 \quad (5.10)$$

Homolysis:

$$H_2O_2 + h\nu \rightarrow 2HO^{\bullet} \quad (5.11)$$

Photodegradation by $^{\bullet}OH$:

$$PH + \bullet OH \rightarrow \bullet P' + H_2O \quad (5.12)$$

5.5 Limitations of TiO_2

The application of TiO_2 nanoparticles in industrial or large-scale water treatment is rather restricted due to its inherent limitations despite being accepted as one of the best semiconductors.[17] The major factors that damp the application of TiO_2 in water treatment applications will be discussed. One of them is the fast recombination rate of photogenerated electron–hole pairs, which lowers the photocatalytic efficiency of TiO_2.[5,18] The photoexcited electron plays a critical role in the generation of reactive oxygen species such as $^{\bullet}O_2^-$, O_2, $^{\bullet}OH$, H_2O_2, and $^{\bullet}O_2H$ by reacting with surface-adsorbed H_2O, OH^-, and O_2.[5] The high rate of recombination indicates a short lifetime of photogenerated charge carriers which would reduce the quantum efficiency and subsequently lower the photocatalytic activity of TiO_2.[19] The wide band gap (approximately 3.2 eV for anatase TiO_2) of TiO_2 limits its operation under UV light only ($\lambda \leq 388$ nm).[7] As a result, TiO_2 can only utilize 3–5% of solar energy from the UV light region ($\lambda_{range} = 290$–400 nm) compared to the larger 45% of the visible light region ($\lambda_{range} = 400$–700 nm)[5,18] limiting its potential to be part of an environmentally sustainable technology.[7] The post-recovery process of TiO_2 nanoparticles in a slurry system is another drawback of TiO_2 photocatalyst.[7,20] The highly dispersed nano-sized TiO_2 particles cause difficulty in the separation of TiO_2 photocatalysts after treatment.[21] As a consequence, water treatment using this technology becomes more costly and time-consuming as additional steps need to be implemented to separate the TiO_2 photocatalyst from the treated water.[22]

5.6 Strategies for Improving Limitations of TiO_2

With the ultimate aim to overcome these limitations, various strategies have been attempted. The strategies include modification of TiO_2 by doping with metals and non-metals,[23–25] coupling with other semiconductors,[26] immobilization on support structures (*e.g.* mesoporous clays, activated carbon, nanofibers, membranes),[27] incorporation of magnetic core (Fe_2O_3 and Fe_3O_4)[28,29] and surface organic modification (*e.g.* dye sensitization).[30]

Doping can be defined as an introduction of foreign atoms into the intrinsic crystal lattice of a semiconductor to impart changes in its electronic properties.[7] By introducing foreign atoms, changes in the optical properties and structural characteristics of a semiconductor can be also attained.[31] Doping has been reported to extend the optical absorption of TiO_2 towards longer wavelengths,[32] and suppress the recombination rate of charge carrier by modifying the band gap.[33] The changes achieved upon doping can enhance the photocatalytic activity and versatility of TiO_2 when applied under visible and UV regions.[34] Both non-metals (*e.g.* N, C, B, S, P, F, I, Cl, Br) and metals (*e.g.* alkaline earth metals, transition metals, and rare earth metals) have been used as dopants to modify the electrical and optical properties of TiO_2.[35–37] Both non-metal and metal doping have successfully imparted changes in the characteristics of the resulting doped products such as band gap, crystal structure, and surface area.

High ionization energy and high electronegativity of non-metal elements such as B, C, F, N, and S enhanced the TiO_2 photocatalytic activity by reducing the band gap and shifting the adsorption edge towards visible light.[38] The band gap reduction is regarded as one of the most critical factors in improving the photocatalytic efficiency of doped TiO_2. Habibi and Jamshidi[39] reported that MB removal under solar light irradiation increased from 37% to 65% upon doping with C due to the band gap reduction from 2.96 eV (419 nm) to 2.71 eV (457.6 nm). The interaction of non-metal element and O 2p electron forms an impurity level above the valence band which extends the optical absorption of TiO_2.[38]

Among the non-metal elements, TiO_2 doped with C or N was found to exhibit superior photocatalytic activity under visible light irradiation.[34] C doping is widely applied as a sensitizer to induce visible light response and to provide stabilization on the anatase phase.[7] Li *et al.*[40] reported that only anatase polymorph was detected after TiO_2 P25 (20% rutile, 80% anatase) was doped with C. Enhanced photocatalytic activity of N-doped TiO_2 under visible light was due to the TiO_2 band gap narrowing.[5] The three notions on band gap narrowing by N dopant are the formation of hybrid N 2p states with O 2p states, occupation of N species on substitutional positions of NO and NO_2[7] or introduction of gap states above the valence band of TiO_2.[17] Rajoriya *et al.*[41] demonstrated that N doping was able to reduce the band gap of TiO_2 from 3.19 eV to 2.98 eV.

In addition, metal doping has also been widely studied.[42] TiO_2 doping with metal involves overlapping of d orbitals of metal ions with the d orbitals of Ti which would aid in increasing absorption intensity and shift the absorption spectrum to longer wavelengths.[42,43] The improvement of TiO_2 photocatalytic activity can be attributed to the extension of the response spectrum range of light absorption and suppression of recombination of photogenerated charge carriers initiated by metal doping.[44] A much more apparent red shift absorption was observed in metal-doped TiO_2, in comparison with non-metals due to the sub-band gap energies of metal elements.[38] For example, Pava-Gómez *et al.*[45] demonstrated a marked redshift absorption from 387.5 nm (3.2 eV) to 529.91 nm (2.34 eV) upon Cu doping. The metal dopants were also able to provide more trapping sites for electrons and holes compared to the non-metal dopants[38] due to the ability of the former to serve as an electron trap when it comes in contact with a semiconductor.[46]

Coupling of TiO_2 with other semiconductors or narrow band gap semiconductors has also successfully produced heterojunction materials that can harvest visible light and separate the photoinduced electrons and holes across the interface.[47]

5.7 Graphitic Carbon Nitride

Graphitic carbon nitride (gC_3N_4) is also known as a metal-free polymeric conductor which consists of carbon and nitrogen.[48] gC_3N_4 was first discovered in 2009 as a promising photocatalyst for hydrogen production by Wang *et al.*[49] Since then, gC_3N_4 has been extensively applied in many applications such as environmental,[50] biomedical,[51] optoelectronic,[52] and energy storage.[53] gC_3N_4 contains seven different phases which were predicted using calculations of the first-principle namely g-h-heptazine, pseudocubic C_3N_4, g-otriazine, g-h-triazine, α-C_3N_4, β-C_3N_4, and cubic C_3N_4 which exhibits band gaps of 2.88 eV, 4.13 eV, 0.93 eV, 2.97 eV, 5.5 eV, 4.85 eV, and 4.3 eV respectively.[54] The heptazine ring structure and high condensation degree allow the metal-free gC_3N_4 to have good chemical stability, high surface area, and an appealing electrical structure with a low band gap of 2.7–2.8 eV.[55] The low energy band gap offers a versatile pathway for attaining control over the

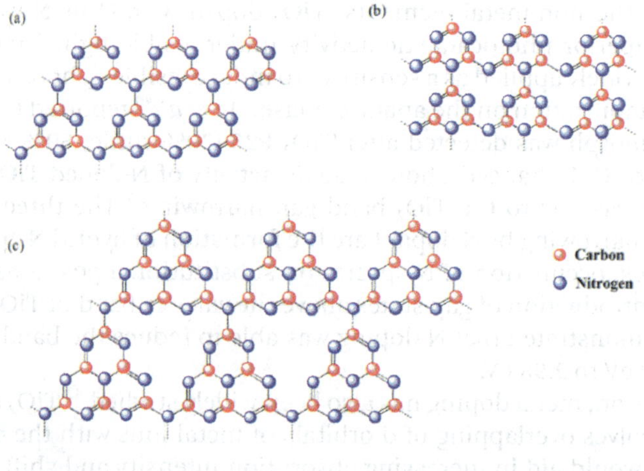

Figure 5.2 (a) *s*-Tris-azine based hexagonal structure, (b) *s*-triazine based orthorhombic structure, (c) tri-*s*-triazine-based structure. Reproduced from ref. 56 with permission from Elsevier, Copyright 2022.

HOMO (highest occupied molecular orbital) and LUMO (lowest unoccupied molecular orbital).[56] Figure 5.2 shows the three forms of gC_3N_4 namely *s*-triazine-bases hexagonal structure, tri-*s*-triazine (heptazine)-based structure, and *s*-triazine-based orthorhombic structure.[56]

gC_3N_4 is highly preferred compared to other carbonaceous materials due to its narrow or mild band gap (~2.7 eV), which makes it highly responsive to visible light illumination contrary to graphene oxide with a larger band gap of 3.0–3.5 eV. The properties of gC_3N_4 lead to a visible-driven photocatalyst for photocatalytic applications utilizing longer wave radiation ($\lambda > 400$ nm).[57] Other carbonaceous materials for example graphene and activated carbon, have low degradation rates due to their low charge mobility and inefficient electron transfer.[58–60] Pure g-C_3N_4 has a conduction band potential of around 1.3 V (normal hydrogen electrode (NHE) at pH = 7), and a valence band position of +1.4 V (NHE at pH 7).[61] gC_3N_4 has high stability and can withstand high temperatures which makes it an excellent candidate for the use of photocatalytic degradation of pollutants in wastewater. It also possesses high charge mobility which exhibits high photocatalytic activity. This is due to its unique electronic structure, which enables it to efficiently generate electron–hole pairs and facilitate redox reactions.

Many precursors can be used to synthesize gC_3N_4 by pyrolysis methods such as cyanamide,[62] melamine,[63] thiourea,[64] and urea.[62] Figure 5.3 illustrates the use of different precursors to synthesize gC_3N_4.[65] Among all the precursors used to synthesize gC_3N_4, urea is a low-cost and widely available precursor, making it a more economically viable option and having a higher nitrogen content.[66] Therefore, the use of urea as a precursor for synthesizing gC_3N_4 offers a cost-effective and efficient approach to producing high-quality carbon materials.

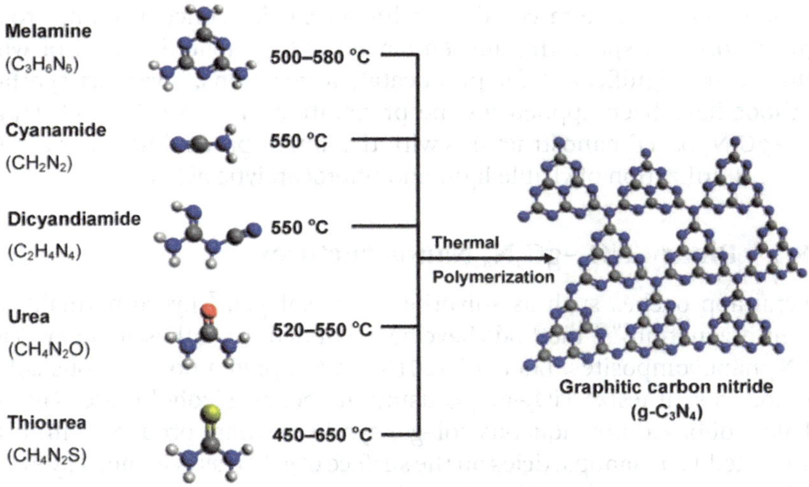

Figure 5.3 Synthesis process of gC_3N_4 from various precursors.[65] Reproduced from ref. 65 with permission from American Chemical Society, Copyright 2016.

The temperature used to synthesize gC_3N_4 plays a crucial role in determining the properties and quality of the resulting material.[67] Typically, the synthesis process involves heating the precursor used at a high temperature in an inert atmosphere, which leads to its decomposition and subsequent carbonization. The temperature range used for the synthesis process is between 500 °C and 800 °C, with the optimum temperature depending on several factors such as the precursor type, reaction time, and heating rate.[68] At lower temperatures, incomplete carbonization can result in the formation of amorphous carbon, while higher temperatures can lead to the formation of graphite. Therefore, the temperature used for synthesizing gC_3N_4 from urea needs to be carefully optimized to achieve the desired properties and quality of the resulting material.

Pristine gC_3N_4 has poor electrical characteristics, including a high rate of electron and hole recombination, low product yield, and low visible light utilization efficiency.[69,70] Combination of gC_3N_4 with other semiconductors, for example, ZnO/gC_3N_4,[71] $gC_3N_4/TiO_2/GO$[72] and Ag/gC_3N_4[73] could reduce the electron–hole recombination, enhance the visible light efficiency and improve the photoactivity of gC_3N_4.[74]

5.8 TiO_2–gC_3N_4-based Nanostructures: Synthesis Methods and Properties

Compared to single photocatalysts, combined binary TiO_2–gC_3N_4 nanocomposites or heterostructure could effectively prevent the recombination of electron–hole pairs. Ternary heterojunctions which typically consist of the multicomponent system could offer higher carrier concentration, broader light absorption spectrum, and enhanced redox capabilities, all of which hold greater significance for photocatalytic reactions.[75] Various synthesis methods have been applied for the preparation of binary TiO_2–gC_3N_4 and TiO_2–gC_3N_4-based nanostructures with the aim of producing materials with excellent utilization of visible light and photocatalytic activity.

5.8.1 Binary TiO_2–gC_3N_4 Nanostructures

Several approaches such as solvothermal,[76] sol–gel,[77] hydrothermal,[78] and wet-impregnation[79,80] methods have been applied to synthesize binary TiO_2–gC_3N_4 nanocomposites. Liu *et al.* (2016)[77] attempted a non-aqueous sol–gel method to synthesize TiO_2–gC_3N_4 using the benzyl-alcohol route. The benzyl-alcohol-based non-aqueous sol–gel synthesis route produced uniformly distributed TiO_2 nanoparticles on the surface of gC_3N_4 as evidenced by FESEM analysis suggesting that this approach successfully produced material with controlled morphology and excellent photoactivity under sunlight. A simple combined sol–gel and hydrothermal method using a low temperature of 300 °C was explored as a route to prepare the binary TiO_2–gC_3N_4 nanocomposite.[81] This synthesis route resulted in a material with absorption towards longer wavelengths and a large specific surface area of 135.6 $m^2\ g^{-1}$ compared to 89.1 $m^2\ g^{-1}$ for gC_3N_4 and 114.3 $m^2\ g^{-1}$ for gC_3N_4.

Hossain *et al.*[82] synthesized TiO_2–gC_3N_4 using melamine and anatase TiO_2 (NP400) as the precursors *via* a facile template-free calcination approach. The material was found to present a narrowed band gap with values ranging from 2.66 eV to 2.81 eV depending on the mixing ratio of NP400 to melamine used in the preparation of the composite. The combination of NP400 and gC_3N_4 in a heterojunction significantly extended the absorption range of the composites, encompassing both the UV and visible regions. Another study also applied the calcination approach to preparing TiO_2–gC_3N_4 photocatalysts using urea and TiO_2 P25 Degussa as precursors.[83] Likewise, the visible-light responsive nanocomposite displayed excellent separation of electron–hole pairs as a result of the heterojunction formation in the photocatalyst.

A facile wet chemical method was applied to synthesize TiO_2–gC_3N_4 nanocomposites using dicyandiamide and titanium butoxide as precursors.[84] This study combined both bulk gC_3N_4 and gC_3N_4 nanosheets to prepare the nanocomposite with the aim of enhancing photocatalytic activities under visible light. It was noted that the nanocomposites (*e.g.* 2T/CN, 4TCN, 6T/CN and 8T/CN) demonstrated higher charge carrier separation compared to gC_3N_4 alone based on the photoluminescence (PL) emission and surface photovoltage spectra response.

A recent study by Matias *et al.*[85] explored a microwave-assisted approach in preparing a visible light-responsive TiO_2–gC_3N_4 photocatalyst. The advantages of the microwave technique in material synthesis include short reaction time, unique heating mechanism, and energy efficiency.[86] The structural and optical properties analyzed *via* XRD and SEM analyses revealed the co-existence of both materials and in close contact to boost photocatalytic efficiency and improve the charge separation. The optimized material, 30 wt% gC_3N_4-TiO_2 successfully removed 85% of methyl orange within 4 hours under simulated solar light.

5.8.2 TiO_2–gC_3N_4-based Nanostructures

The hydrothermal method has been a popular choice in the fabrication of TiO_2–gC_3N_4-based nanostructures. Various studies have embarked on this method primarily due to its ability to provide controlled and reproducible fabrication, homogeneous mixing of precursors, high reaction efficiency, low energy consumption, versatility, scalability, and the generation of high-surface-area nanocomposites. A facile hydrothermal procedure was applied for the synthesis of titanium dioxide/zirconium dioxide/graphitic carbon nitride (TiO_2/ZrO_2/gC_3N_4, TZCN) ternary composites whereby melamine and tetrabutyl titanate were used as precursors for the material preparation.[87] The optimized sample, TZCN-1 possessed low emission intensity suggesting the ability to effectively separate electron–hole pairs and the smallest band gap of 2.48 eV and the highest surface area of 188.65 $m^2\ g^{-1}$. All these properties were attributed to the formation of a heterojunction, which facilitated the movement, prolonged the existence, and improved the isolation of photoinduced charge carriers.

A Z-scheme ternary nanocomposite of Fe_3O_4/TiO_2/gC_3N_4 was prepared through a novel and *in situ* growth mechanism based on the hydrothermal route, employing melamine, tetrabutyl titanate, and pre-prepared magnetic Fe_3O_4 nanoparticles.[88] Although a slight decrease in surface area was observed in the Fe_3O_4/TiO_2/gC_3N_4 material after Fe_3O_4 particles were loaded onto TiO_2/gC_3N_4 (72.58 $m^2\ g^{-1}$ to 65.74 $m^2\ g^{-1}$), it was revealed by UV-vis diffuse reflectance results that the photocatalytic activity of g-C_3N_4 was enhanced by the addition of TiO_2 and Fe_3O_4. This enhancement was achieved by shifting the band threshold of Fe_3O_4/TiO_2/gC_3N_4 into the visible region (610 nm), narrowing the band gap to 1.9 eV, and extending the absorption of light into the visible spectrum.

SrO doped on mesoporous gC_3N_4/TiO_2 was conducted *via* a two-step protocol consisting of *in situ* sol-precipitation and hydrothermal treatment.[89] The photoluminescence analysis revealed that the Z-scheme heterojunction photocatalyst, SrO-mpg-gC_3N_4/TiO_2 possessed great charge separation after the introduction of SrO to the surface of mpg-gC_3N_4/TiO_2. On the other hand, optical properties based on UV–Vis-NIR diffuse reflectance spectroscopy showed that the band gap energy of SrO-mpg-gC_3N_4/TiO_2 increased to 3.22 eV compared to that of mpg-gC_3N_4 and mpg-gC_3N_4/TiO_2 with 2.82 eV and 3.08 eV, respectively. The study concluded that although the band gap increased, the recombination of electron–hole pairs reduced in the synthesized catalysts. This reduction can be attributed to the creation of heterojunctions between mpg-gC_3N_4 and TiO_2 semiconductors. A recent study by Sewnet *et al.*[90] also applied a hydrothermal route for the preparation of ternary g-C_3N_4/Ag–TiO_2 heterostructures using urea, thiourea, and titanium isopropoxide as precursors. The remarkably enhanced photocatalytic performance was attributed to the combined influence of Ag's localized surface plasmon resonance (LSPR) effects, its function as an electron-conduction bridge, and the presence of a highly effective heterostructure between Ag–TiO_2 and g-C_3N_4.

5.9 Applications of TiO_2–gC_3N_4-based Nanocomposites for Emerging Pollutant Removal

Pharmaceuticals and personal care products (PPCPs) are relatively recent categories of emerging environmental pollutants that have gained significant attention due to their environmental effects, which include their frequent presence, persistence, and potential risks to both aquatic ecosystems and human populations. PPCPs are widely present globally, including within drinking water treatment facilities. Among these contaminants, antibiotics in water and waterborne bacteria are particularly concerning, given their widespread presence and the adverse biological and ecotoxicological effects they can pose. Wastewater from sources like animal farming, aquaculture, and the pharmaceutical industry is a major contributor to environmental antibiotics.

One group of antibiotics, known as fluoroquinolones (FQs), was introduced in the 1970s and has several advantages, including a broad range of antibacterial capabilities and minimal side effects. Among them, ofloxacin, ciprofloxacin (CIP), and enrofloxacin (ENX) are widely used. Conventional wastewater treatment plants are inefficient in treating emerging pollutants resulting in their detection as well as metabolites in the environmental components.[91] TiO_2–gC_3N_4-based nanocomposites, when integrated into wastewater treatment systems, can efficiently degrade emerging pollutants and ensure the release of treated effluent meeting stringent environmental standards. This application not only contributes to cleaner water bodies but also promotes sustainable industrial practices. For instance, Su *et al.*[92] developed a novel photocatalyst called CDs/TiO_2/gC_3N_4, which is driven by visible light for the removal of enrofloxacin. They achieved this by incorporating carbon dots (CDs) into TiO_2 nanoparticles and coating the surface of gC_3N_4 nanosheets through a simple hydrothermal process. This photocatalyst demonstrated exceptional performance, with 91.6% decomposition of ENX within just 1 hour. This efficiency was approximately 3.95 times higher than that of TiO_2, 4.82 times higher than gC_3N_4, and 1.69 times higher than TiO_2 (90.0 wt%)/gC_3N_4. The enhanced photocatalytic activity was attributed to the unique up-converted photoluminescent property of CDs, efficient charge separation, and effective electron transportation.

CIP can encourage the development of antibiotic resistance genes where they can lead to the proliferation of antibiotic-resistant bacteria, posing a potential threat to both human health and the balance of microbial communities in the environment. In a study by Li *et al.*,[93] a 3D composite material consisting of gC_3N_4, TiO_2, and kaolinite was fabricated through a mild sol–gel procedure, involving chemical stripping and self-assembly techniques. This composite demonstrated enhanced efficacy in the photodegradation of CIP when exposed to visible light. The research findings elucidated the formation of a distinctive "sandwich" structure within the composite. This structure facilitated the uniform dispersion of nano TiO_2 particles with reduced grain sizes, increased exposure of gC_3N_4 sheets, an enlarged surface area, and increased visible-light absorption capacity. Notably, the composite exhibited an apparent rate constant approximately 5.35 times greater than pure TiO_2, 6.35 times greater than gC_3N_4 alone, and 4.49 times greater than the widely-used P25 material in the context of CIP removal through photodegradation.

Tetracycline hydrochloride (TCH), formed through the combination of hydrochloric acid and tetracycline, possesses increased water solubility. This compound can endure in a water-based environment for an extended period, serving as a persistent antibiotic substance. A reusable membrane, made from a combination of gC_3N_4@TiO_2 nanowires and modified graphene oxide (referred to as GO/CN@TNWs), was successfully produced using electrostatic interactions and a vacuum filtration-assisted self-assembly process.[94] This newly created membrane exhibited significantly higher removal of TCH when exposed to simulated solar light attributed to the presence of GO, which has strong light absorption and carrier mobility properties. The p–p stacking played a significant role in enhancing the removal rate of 98.77% for TCH. In-depth studies using room-temperature electron spin resonance (ESR) demonstrated that the GO/CN@TNWs membrane could generate higher quantities of superoxide radicals, hydroxyl radicals, and hole species, all of which are crucial for driving the photocatalytic degradation process.

Ibuprofen (IBU), a non-steroidal anti-inflammatory drug (NSAID) produced at a global scale of several thousand tons annually, is commonly utilized for pain relief, fever reduction, and inflammation management. Due to its resistant and persistent nature, traditional wastewater treatment plants face challenges in efficiently removing IBU. Consequently, IBU is frequently found in treated wastewater at concentrations ranging from 20 ng L^{-1} to 3.8 μg L^{-1}.[95] For the removal of ibuprofen, a highly efficient and recyclable photocatalyst called $gC_3N_4/TiO_2/Fe_3O_4$@SiO_2 (gCTFS) through a sol–gel method was developed using only a small number of ultrathin gC_3N_4 nanosheets.[96] This nanophotocatalyst exhibited exceptional photoactivity, removing 97% of ibuprofen in just 15 minutes of visible light exposure. Furthermore, the nanophotocatalyst exhibited superparamagnetic behavior with a saturation magnetization of 8 emu g^{-1}, allowing for easy magnetic recovery from the solution for reuse in subsequent applications.

Acetaminophen (APAP), also known as paracetamol, is a widely employed NSAID with a prominent role in the management of conditions such as colds, fever, headaches, and nerve pain. gC_3N_4@TiO_2 (gT) and gC_3N_4@TiO_2@ZnO (gTZ) prepared *via* the sol–gel technique and sol–gel heat condensation process demonstrated remarkable efficacy in breaking down APAP.[97] Specifically, gT achieved a substantial degradation rate of 70.6%, whereas gTZ exhibited superior performance with an impressive degradation rate of 92.4%. The primary agents responsible for APAP degradation in the case of gTZ were hydroxyl radicals ($^{\bullet}OH$).

Several studies have also reported the application of TiO_2–gC_3N_4-based nanocomposites for the removal of pesticides. For example, Nekooie *et al.*[98] reported that $gC_3N_4/Cu/TiO_2$ composite successfully removed 60% of endosulfan, a toxic organochlorine pesticide after 80 min of irradiation under visible light. High removal of imidacloprid pesticide (93%) was attained using gC_3N_4/TiO_2 composites. Another study reported high efficacy of TiO_2–gC_3N_4 nanocomposite in removing 2-chlorophenol and 4-chlorophenol as a result of the decrease in band gap energy and the inhibition of charge carriers on the surfaces of both TiO_2 and gC_3N_4.[99]

5.10 Conclusion

The development of TiO_2–graphitic carbon nitride-based nanocomposites for the degradation of emerging pollutants represents a significant advancement in the field of photocatalysis and environmental remediation. Through synergistic integration of TiO_2 and graphitic carbon nitride, these nanocomposites exhibit enhanced photocatalytic activity, extended light absorption range, and improved charge separation efficiency, thereby offering promising solutions for addressing the challenges posed by emerging pollutants in the environment.

The exploration of advanced synthesis techniques and innovative approaches for controlling the structural and physicochemical properties of binary and ternary TiO_2–graphitic carbon nitride-based nanocomposites holds immense potential for further improving their efficiency and scalability in practical applications. Strategies such as doping, surface modification, and hybridization with other nanomaterials offer avenues for fine-tuning the photocatalytic properties and addressing the inherent limitations of these nanocomposites.

The future outlook involves advancing nanocomposite design, developing efficient photocatalytic reactors, implementing environmental monitoring and regulation, prioritizing sustainability, raising public awareness, conducting demonstration projects, and promoting global collaboration. These efforts are essential for ensuring clean and safe water resources and mitigating the impact of emerging pollutants on the environment and human health.

References

1. O. Carp, C. L. Huisman and A. Reller, *Prog. Solid State Chem.*, 2004, **32**, 33–177.
2. H. Siddiqui, *Ion Beam Tech. Appl.*, 2019, **83566**, 1–13.

3. C. Luo, X. Ren, Z. Dai, Y. Zhang, X. Qi and C. Pan, *ACS Appl. Mater. Interfaces*, 2017, **9**, 23265–23286.
4. A. Di Paola, M. Bellardita and L. Palmisano, *Catalysts*, 2013, **3**, 36–73.
5. V. Etacheri, C. Di Valentin, J. Schneider, D. Bahnemann and S. C. Pillai, *J. Photochem. Photobiol., C*, 2015, **25**, 1–29.
6. S. M. Gupta and M. Tripathi, *Chin. Sci. Bull.*, 2011, **56**, 1639–1657.
7. A. T. Kuvarega and B. B. Mamba, *Crit. Rev. Solid State Mater. Sci.*, 2017, **42**, 295–346.
8. N. A. A. Aziz, P. Palaniandy, H. A. Aziz and I. Dahlan, *J. Chem. Res.*, 2016, **40**, 704–712.
9. J. Zhang, P. Zhou, J. Liu and J. Yu, *Phys. Chem. Chem. Phys.*, 2014, **16**, 20382–20386.
10. A. Bakri, M. Sahdan, F. Adriyanto, N. Raship, N. Said, S. Abdullah and M. Rahim, *AIP Conf. Proc.*, 2017, **1788**, 030030.
11. Y. Luo, A. Benali, L. Shulenburger, J. T. Krogel, O. Heinonen and P. R. Kent, *New J. Phys.*, 2016, **18**, 113049.
12. D. A. Hanaor and C. C. Sorrell, *J. Mater. Sci.*, 2011, **46**, 855–874.
13. M. N. Chong, B. Jin, C. W. Chow and C. Saint, *Water Res.*, 2010, **44**, 2997–3027.
14. M. Pawar, S. Topcu Sendoğdular and P. Gouma, *J. Nanomater.*, 2018, 2018.
15. S. Martha, P. C. Sahoo and K. Parida, *RSC Adv.*, 2015, **5**, 61535–61553.
16. X. Kang, S. Liu, Z. Dai, Y. He, X. Song and Z. Tan, *Catalysts*, 2019, **9**, 191.
17. V. Kumaravel, S. Mathew, J. Bartlett and S. C. Pillai, *Appl. Catal., B*, 2018, **244**, 1021–1064.
18. X. Lei, X. Xue and H. Yang, *Appl. Surf. Sci.*, 2014, **321**, 396–403.
19. A. Bumajdad and M. Madkour, *Phys. Chem. Chem. Phys.*, 2014, **16**, 7146–7158.
20. D. Awfa, M. Ateia, M. Fujii, M. S. Johnson and C. Yoshimura, *Water Res.*, 2018, **142**, 26–45.
21. H. Dong, G. Zeng, L. Tang, C. Fan, C. Zhang, X. He and Y. He, *Water Res.*, 2015, **79**, 128–146.
22. D. Kanakaraju, Y.-C. Lim and A. Pace, *Sustainable Chem. Pharm.*, 2018, **8**, 50–62.
23. D. He, Y. Li, J. Wu, Y. Yang and Q. An, *Appl. Surf. Sci.*, 2017, **391**, 318–325.
24. H. Khan, Z. Jiang and D. Berk, *Sol. Energy*, 2018, **162**, 420–430.
25. E. Bayan, T. Lupeiko, L. Pustovaya, M. Volkova, V. Butova and A. Guda, *J. Alloys Compd.*, 2020, **822**, 153662.
26. A. Hamrouni, H. Azzouzi, A. Rayes, L. Palmisano, R. Ceccato and F. Parrino, *Nanomaterials*, 2020, **10**, 795.
27. F. C. S. Lopes, C. Maria da Graça, P. Bargiela, H. S. Ferreira and C. A. d. M. Pires, *Chem. Eng. Sci.*, 2020, **227**, 115939.
28. M. Behpour, P. Shirazi and M. Rahbar, *React. Kinet., Mech. Catal.*, 2019, 1–13.
29. Q. Sun, Y. Hong, Q. Liu and L. Dong, *Appl. Surf. Sci.*, 2018, **430**, 399–406.
30. L. Kavan, Z. Vlckova Zivcova, M. Zlamalova, S. M. Zakeeruddin and M. Grätzel, *J. Phys. Chem. C*, 2020, **124**, 6512–6521.

31. M. R. Panigrahi and M. Devi, *J. Phys.: Conf. Ser.*, 2019, **1172**, 012046.
32. K. Wetchakun, N. Wetchakun and S. Sakulsermsuk, *J. Ind. Eng. Chem.*, 2018, **71**, 19–49.
33. W. Xie, R. Li and Q. Xu, *Sci. Rep.*, 2018, **8**, 8752.
34. R. Daghrir, P. Drogui and D. Robert, *Ind. Eng. Chem. Res.*, 2013, **52**, 3581–3599.
35. S. Z. Islam, S. Nagpure, D. Y. Kim and S. E. Rankin, *Inorganics*, 2017, **5**, 15.
36. K. Xie, Q. Jia, Y. Wang, W. Zhang and J. Xu, *Materials*, 2018, **11**, 179.
37. A. Dorothy, G. Subramaniam and P. Panigrahi, *Mater. Res. Express*, 2019, **6**, 045913.
38. F. Huang, A. Yan and H. Zhao, *Semicond. Photocatal.: Mater., Mech. Appl.*, 2016, 31–80.
39. S. Habibi and M. Jamshidi, *Environ. Technol.*, 2020, **41**, 3233–3247.
40. W. Li, R. Liang, N. Y. Zhou and Z. Pan, *ACS Omega*, 2020, **5**, 10042–10051.
41. S. Rajoriya, S. Bargole, S. George, V. K. Saharan, P. R. Gogate and A. B. Pandit, *Sep. Purif. Technol.*, 2019, **209**, 254–269.
42. J. Moma and J. Baloyi, *Photocatalysts*, 2019, **79374**, 2–21.
43. O. Ola and M. M. Maroto-Valer, *Appl. Catal., A*, 2015, **502**, 114–121.
44. F. Zhang, X. Wang, H. Liu, C. Liu, Y. Wan, Y. Long and Z. Cai, *Appl. Sci.*, 2019, **9**, 2489.
45. B. Pava-Gómez, X. Vargas-Ramírez and C. Díaz-Uribe, *J. Photochem. Photobiol., A*, 2018, **360**, 13–25.
46. A. Amorós-Pérez, L. Cano-Casanova, A. Castillo-Deltell, M. Lillo-Ródenas and M. Román-Martínez, *Materials*, 2019, **12**, 40.
47. Y. Zhao, X. Linghu, Y. Shu, J. Zhang, Z. Chen, Y. Wu, D. Shan and B. Wang, *J. Environ. Chem. Eng.*, 2022, **10**, 108077.
48. G. Mamba and A. Mishra, *Appl. Catal., B*, 2016, **198**, 347–377.
49. X. Wang, K. Maeda, X. Chen, K. Takanabe, K. Domen, Y. Hou, X. Fu and M. Antonietti, *J. Am. Chem. Soc.*, 2009, **131**, 1680–1681.
50. Q. Hao, G. Jia, W. Wei, A. Vinu, Y. Wang, H. Arandiyan and B.-J. Ni, *Nano Res.*, 2020, **13**, 18–37.
51. G. Liao, F. He, Q. Li, L. Zhong, R. Zhao, H. Che, H. Gao and B. Fang, *Prog. Mater. Sci.*, 2020, **112**, 100666.
52. J. Bian, C. Huang and R. Q. Zhang, *ChemSusChem*, 2016, **9**, 2723–2735.
53. W. Zhao, Z. Yan and L. Qian, *Eng. Sci.*, 2020, **10**, 24–34.
54. C. Prasad, H. Tang and I. Bahadur, *J. Mol. Liq.*, 2019, **281**, 634–654.
55. Z. Zhao, Y. Sun and F. Dong, *Nanoscale*, 2015, **7**, 15–37.
56. D. Vaya, B. Kaushik and P. K. Surolia, *Mater. Sci. Semicond. Process.*, 2022, **137**, 106181.
57. G. Liao, Y. Gong, L. Zhang, H. Gao, G.-J. Yang and B. Fang, *Energy Environ. Sci.*, 2019, **12**, 2080–2147.
58. J. Ge, Y. Zhang and S.-J. Park, *Materials*, 2019, **12**, 1916.
59. N. Khalid, A. Majid, M. B. Tahir, N. Niaz and S. Khalid, *Ceram. Int.*, 2017, **43**, 14552–14571.
60. Y. Zhang, K. Hawboldt, L. Zhang, J. Lu, L. Chang and A. Dwyer, *Appl. Catal., A*, 2022, **630**, 118460.

61. A. Sewnet, M. Abebe, P. Asaithambi and E. Alemayehu, *Air, Soil Water Res.*, 2022, **15**, 11786221221117266.
62. J.-G. Kim, S.-M. Park, M. E. Lee, E. E. Kwon and K. Baek, *Chemosphere*, 2018, **212**, 193–199.
63. M. Bellardita, E. I. García-López, G. Marcì, I. Krivtsov, J. R. García and L. Palmisano, *Appl. Catal., B*, 2018, **220**, 222–233.
64. S. Hua, D. Qu, L. An, W. Jiang, Y. Wen, X. Wang and Z. Sun, *Appl. Catal., B*, 2019, **240**, 253–261.
65. W.-J. Ong, L.-L. Tan, Y. H. Ng, S.-T. Yong and S.-P. Chai, *Chem. Rev.*, 2016, **116**, 7159–7329.
66. S. J. Mun and S.-J. Park, *Catalysts*, 2019, **9**, 805.
67. D. R. Paul, R. Sharma, S. Nehra and A. Sharma, *RSC Adv.*, 2019, **9**, 15381–15391.
68. K. Qi, S.-y. Liu and A. Zada, *J. Taiwan Inst. Chem. Eng.*, 2020, **109**, 111–123.
69. K. Li, F.-Y. Su and W.-D. Zhang, *Appl. Surf. Sci.*, 2016, **375**, 110–117.
70. C. Ye, J.-X. Li, Z.-J. Li, X.-B. Li, X.-B. Fan, L.-P. Zhang, B. Chen, C.-H. Tung and L.-Z. Wu, *ACS Catal.*, 2015, **5**, 6973–6979.
71. A. Bhosale, A. Gophane, J. Kadam, S. Sabale, K. Sonawane and K. Garadkar, *Opt. Mater.*, 2023, **136**, 113392.
72. P. Lin, H. H. Hu, H. R. Lv, Z. X. Ding, L. B. Xu, D. G. Qian, P. Wang, J. Q. Pan, C. R. Li and C. Cui, *Appl. Phys. A: Mater. Sci. Process.*, 2018, **124**, 11.
73. P. C. Nagajyothi, M. Pandurangan, S. V. P. Vattikuti, C. O. Tettey, T. V. M. Sreekanth and J. Shim, *Sep. Purif. Technol.*, 2017, **188**, 228–237.
74. B. C. Zhu, L. Y. Zhang, B. Cheng and J. G. Yu, *Appl. Catal., B*, 2018, **224**, 983–999.
75. H. Liu, F. Sun, X. Li, Q. Ma, G. Liu, H. Yu, W. Yu, X. Dong and Z. Su, *Composites, Part B*, 2023, **259**, 110746.
76. D. Lu, P. Fang, W. Wu, J. Ding, L. Jiang, X. Zhao, C. Li, M. Yang, Y. Li and D. Wang, *Nanoscale*, 2017, **9**, 3231–3245.
77. X. Liu, N. Chen, Y. Li, D. Deng, X. Xing and Y. Wang, *Sci. Rep.*, 2016, **6**, 39531.
78. R. Mohini and N. Lakshminarasimhan, *Mater. Res. Bull.*, 2016, **76**, 370–375.
79. R. A. Senthil, J. Theerthagiri, A. Selvi and J. Madhavan, *Opt. Mater.*, 2017, **64**, 533–539.
80. A. Kane, L. Chafiq, S. Dalhatou, P. Bonnet, M. Nasr, N. Gaillard, J. M. D. Dikdim, G. Monier, A. A. Assadi and H. Zeghioud, *J. Photochem. Photobiol., A*, 2022, **430**, 113971.
81. M. Kocijan, M. Vukšić, M. Kurtjak, L. Ćurković, D. Vengust and M. Podlogar, *Catalysts*, 2022, **12**, 1554.
82. S. M. Hossain, H. Park, H.-J. Kang, J. S. Mun, L. Tijing, I. Rhee, J.-H. Kim, Y.-S. Jun and H. K. Shon, *Chemosphere*, 2021, **262**, 128004.
83. S.-H. Liu and W.-X. Lin, *J. Photochem. Photobiol., A*, 2019, **378**, 66–73.
84. F. Raziq, C. Li, M. Humayun, Y. Qu, A. Zada, H. Yu and L. Jing, *Mater. Res. Bull.*, 2015, **70**, 494–499.

85. M. L. Matias, A. S. Reis-Machado, J. Rodrigues, T. Calmeiro, J. Deuermeier, A. Pimentel, E. Fortunato, R. Martins and D. Nunes, *Nanomaterials*, 2023, **13**, 1090.
86. R. Krishnan, S. N. Shibu, D. Poelman, A. K. Badyal, A. K. Kunti, H. C. Swart and S. G. Menon, *Mater. Today Commun.*, 2022, **32**, 103890.
87. P. Ding, H. Ji, P. Li, Q. Liu, Y. Wu, M. Guo, Z. Zhou, S. Gao, W. Xu, W. Liu, Q. Wang and S. Chen, *Appl. Catal., B*, 2022, **300**, 120633.
88. A. Raza, H. Shen, A. A. Haidry and S. Cui, *Appl. Surf. Sci.*, 2019, **488**, 887–895.
89. D. Kılıç, M. Sevim, Z. Eroğlu, Ö. Metin and S. Karaca, *Adv. Powder Technol.*, 2021, **32**, 2743–2757.
90. A. Sewnet, E. Alemayehu, M. Abebe, D. Mani, S. Thomas and B. Lennartz, *Materials*, 2023, **16**, 5497.
91. Y. Zhang, X. Zhang and S. Wang, *Sci. Total Environ.*, 2023, **883**, 163702.
92. Y. Su, P. Chen, F. Wang, Q. Zhang, T. Chen, Y. Wang, K. Yao, W. Lv and G. Liu, *RSC Adv.*, 2017, **7**, 34096–34103.
93. C. Li, Z. Sun, W. Zhang, C. Yu and S. Zheng, *Appl. Catal., B*, 2018, **220**, 272–282.
94. Y. Huang, G. Zhu, K. Zou, F. Tian, T. Prasad Yadav, H. Xu, G. Yang, H. Li and L. Qu, *J. Mol. Liq.*, 2021, **337**, 116461.
95. E. N. Evgenidou, I. K. Konstantinou and D. A. Lambropoulou, *Sci. Total Environ.*, 2015, **505**, 905–926.
96. A. Kumar, M. Khan, X. Zeng and I. M. C. Lo, *Chem. Eng. J.*, 2018, **353**, 645–656.
97. Y. Jiang, H. Xi, L. Li, Z. Wang and C. Yang, *J. Photochem. Photobiol., A*, 2023, **445**, 115010.
98. R. Nekooie, J. B. Ghasemi, A. Badiei, T. Shamspur, A. Mostafavi and S. Moradian, *J. Mol. Struct.*, 2022, **1258**, 132650.
99. T. Kobkeatthawin, J. Trakulmututa, T. Amornasakchai and S. M. Smith, *Catalysts*, 2022, **12**, 120.

CHAPTER 6

Carbon-based Nanomaterials for the Removal of Emerging Water Pollutants

WAN HAZMAN DANIAL[a*], NURASYIKIN HAMZAH[a], MOHAMAD WAFIUDDIN ISMAIL[a], NURUL IMAN AMINUDIN[a], SAIFUL ARIFIN SHAFIEE[a] AND ANWAR IQBAL[b]

[a]Department of Chemistry, Kulliyyah of Science, International Islamic University Malaysia, 25200 Kuantan, Pahang, Malaysia; [b]School of Chemical Sciences, Universiti Sains Malaysia, 11800 Gelugor, Penang, Malaysia
*E-mail: whazman@iium.edu.my

6.1 Introduction

In recent times, emerging pollutants have become a significant global concern due to the potential risks they pose to both the environment and human well-being. Emerging pollutants are substances identified in water yet lacking comprehensive testing or regulation, they encompass compounds such as pesticides, pharmaceuticals, hormones and steroids, cosmetics, personal care products (PCPs), endocrine disrupting chemicals (EDCs), perfluoronated compounds (PFCs), microplastics and industrial additives. Even in minute concentrations, these substances possess the potential to detrimentally affect both human well-being and the ecosystem.[1] Emerging pollutants found in surface water stem from human or animal sources, either being directly discharged into water bodies or gradually leaching into soils.

Chemistry in the Environment Series No. 14
Advanced Materials for Emerging Water Pollutant Removal
Edited by Pei Sean Goh, Devagi Kanakaraju, Anwar Iqbal and Ahmad Fauzi Ismail

Published by the Royal Society of Chemistry, www.rsc.org

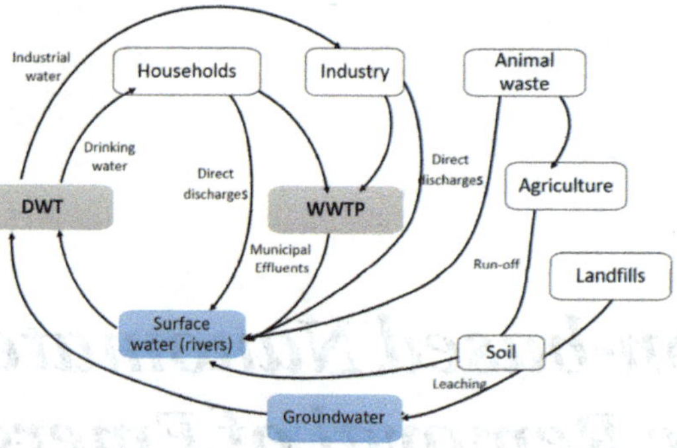

Figure 6.1 Emerging pollutant pathways *via* water route.[3] Reproduced from ref. 3 with permission from Elsevier, Copyright 2018.

The pathway of emerging pollutants. from various sources related to water uses is illustrated in Figure 6.1. This leads to a decline in water quality, contamination of drinking water sources, and the emergence of health-related challenges. Considering that drinking water treatment facilities rely on water resources, the prevalence of such contamination in aquatic ecosystems, particularly surface water, presents a significant and pressing issue.[2] Consequently, removing then is of paramount importance.

Pollutants can be degraded and removed from water bodies *via* treatment methods categorized as biological, physicochemical, and oxidation. Activated sludge treatment, membrane bioreactors, nitrification and denitrification, and microorganism-based treatment are examples of biological treatments for the removal of emerging pollutants. Physicochemical treatment technologies work to remove various emerging pollutants *via* coagulation–flocculation while oxidation methods comprise chlorination, ozone treatment, photocatalysis, Fenton oxidation, and electrooxidation.[4] However, these methods are also associated with their own shortcomings of removal inefficiency, cost effectiveness, and method feasibility and adaptability. Adsorption process, a type of physicochemical method is found to be an economical and viable approach in simultaneous removal of emerging pollutants using a wide range of adsorbents. Generally, adsorbates attach to a solid surface or an interface adsorbent *via* van der Waals and/or electrostatic attractions through a dynamic equilibrium process of physisorption or chemisorption.[5] Presently, there is a growing inclination towards the usage of chemically modified organic or agricultural wastes such as carbon adsorbents for water treatment, primarily driven by their low cost, biodegradable nature, and their straightforward processes. Despite, the promising applicability, the adsorbents are mainly produced as carbon micromaterials having limited accessible surface area leading to poor affinity, low selectivity, and prolonging maximum pollutant removal time.[6] Hence, carbon nanomaterial is comparatively more efficient in overcoming these inherent limitations of micromaterials.

Carbon nanomaterials such as carbon nanoparticles, carbon nanotubes, carbon nanofibers, carbon black, fullerenes (C60), graphene, and nanodiamonds[7] are superb materials for environmental remediation purposes due to their large surface area, good sustainability, high abundance in nature and electrical conductance, customizable structure architecture as well as having remarkable stability under severe conditions.[8] These nanomaterials can be further classified according to their dimensionalities *i.e.* 0D, 1D, 2D and 3D. 0D is a zero-dimensional material such as a carbon quantum dot with all three dimensions in the nanoscale range of 1–100 nm. Meanwhile, 1D is a one-dimensional material that possesses two nanoscale and one microscale, or even larger, external dimensions. Nanofibers, nanotubes, nanowires, and nanorods belong to this dimensionality. Nanosheets, nano-coatings and thin films are 2D nanomaterials with only one nanoscale dimension while nanocomposites or assemblies of nanostructured materials are considered to be 3D nanomaterials having no nanoscale external dimensions.[6]

Carbon nanomaterials with their high surface area, significant chemical reactivity, mechanical properties, cost-effectiveness, and low power consumption, can be meaningfully applied for water treatment and remediation. These materials have the potential to serve as effective adsorptive agents since their particle size and porosity are well-defined, can be engineered and are recyclable.[9] Each carbon nanomaterial has unique properties and advantages for water and wastewater treatment that need to be understood. This chapter underscores the advanced sustainable nanomaterials for the elimination of emerging pollutants and contaminants using carbon-based nanomaterials, namely activated carbon nanoparticles, nanodiamonds, fullerenes, graphene and its oxides, carbon and graphene quantum dots, carbon nanotubes and nanofibers, and carbon aerogel. The chapter aims to provide guidelines to the readers to select the best carbon nanomaterials for the intended purposes.

6.2 Carbon-based Nanomaterials for the Removal of Emerging Water Pollutants

6.2.1 Activated Carbon Nanoparticles

Activated carbon nanoparticles (ACNPs) are processed or synthesized carbonaceous materials that possess uniform surface morphology, exceptional biocompatibility, abundant pores with high surface area and significant chemical reactivity as well as remarkable mechanical strength and stability.[10] ACNPs are also known as non-graphite microcrystals since they are mainly made up of microcrystalline graphite structure. The microcrystalline structure of ACNPs has an interlayer spacing of 0.34–0.35 nm. The pore size diameter can be classified into three types, *i.e.* micropores (<2 nm), mesopores (2–50 nm) and macropores (>5 nm)[11] (see Figure 6.2).

ACNPs can be easily synthesized from any carbonaceous wastes which can further be categorized into two types: (1) degradable or low-valued biowaste

Figure 6.2 ACNPs pore size illustration.[10] Reproduced from ref. 10 with permission from Elsevier, Copyright 2018.

such as grass, leaves, rice husks, coconut shells, bacterial cellulose, and waste paper; and (2) non-degradable or negatively valued wastes including plastics and rubber tyres. There are various methods to synthesis the carbon nanoparticles (CNPs) including solution combustion, chemical vapour deposition, high temperature pyrolysis, and hydrothermal synthesis. The CNPs are further activated by chemical and physical methods to yield ACNPs, predominantly to increase the surface area. The physical activation method is a two-step process starting from carbonization of the precursor material at high temperatures and inert conditions to obtain CNPs followed by activation using gas agents including air, steam, and carbon dioxide. Meanwhile, chemical activation is a one-step process that involves chemical reaction on a carbon precursor material in the presence of chemical agents such as acids, bases, and salts.[12]

To enhance the adsorption capacity of the ACNPs for a higher uptake of specific pollutants, surface modification is usually carried out after the activation step *via* three methods: chemical, physical, and biological modifications. Examples of chemical modification include acidic and basic treatment as well as incorporation of ACNPS with metal oxide, nano-sized zero-valent iron, iron-based nanoparticles, or specific chemical compounds. The physical modification techniques that have been used to modify ACNPs include microwave, ozone, and plasma treatment while the biological modification involves the use of trapped bacteria and immobilized enzymes[13] (see Figure 6.3). ACNPs offer exceptional advantages over other adsorbents in wastewater treatment due to their structural porosity and substantial surface-to-volume ratio. This unique adsorption characteristic is determined by the presence of functional groups which originate from the precursor material, activation type, and other thermal purification techniques. Key functional groups like carboxyl (–COOH), carbonyl (C=O), phenol (C_6H_5OH), and lactones (–(C=O)–O–) play a significant role in adsorbing contaminants in carbon structures, among other groups. Moreover, hydrogen, oxygen, and sulphur are the commonly found atoms within the ACNPs structure.[12]

Pharmaceuticals are regarded as emerging contaminants in water sources, posing a significant toxicological threat to aquatic organisms. The World Health Organization (WHO) has sanctioned the use of ACNP adsorption as an effective technique in water treatment plants to eliminate pharmaceutical compounds, owing to its affordability and straightforward application.

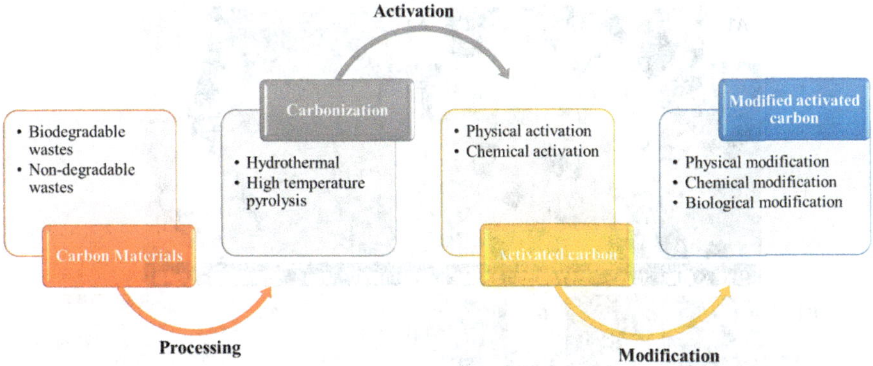

Figure 6.3 General steps for preparation of modified activated carbon.

Some emerging pharmaceutical pollutants contain the structure of the phenolic ring such as salicylic acid and methylparaben. The typical main adsorption interactions between the phenolic compounds on the ACNPs are π-stacking, hydrogen bonds, ionic interactions, and van der Waals forces.[14] Remarkable applicability of ACNPs produced from plant or agricultural wastes to remove antibiotics entering water bodies were majorly reported and the mechanism of removal was mainly *via* adsorption. The adsorption mechanism is universal and is the preferential treatment process due to its simplicity in design and operation with high efficiency and feasibility. For instance, the ACNPs derived from *Azolla filiculoides* fern are able to adsorb ampicillin antibiotic with 96.84% efficiency.[15] Chemical modification of the pore texture and surface area of the ACNPs can lead to an increase in sorption ability. Hashemzadeh *et al.*[16] examined the activity of ACNPs from aloe vera leaf waste activated by sulphuric acid and nitric acid, respectively for removal of amoxicillin and cephalexin in contaminated media. The modified ACNPs noticeably enhanced the adsorption capacity for these two drugs (25.56–29.11 mg g^{-1}) compared to other adsorbents with reasonable results (2.69–17.68 mg g^{-1}).[16] The disappearance or size reduction of pores, larger fragments and flatter surface of ACNPs after water treatment are common observations indicating high volumes of drug molecule attachment and adsorption (see Figure 6.4).

Moreover, apart from being a source of carbon, agricultural waste also can be effectively harnessed as an enzyme immobilization support for the efficient removal of persistent pharmaceutical pollutants. Drugs such as tetracycline antibiotics, carbamazepine, and iopamidol are hard to degrade due to their low biodegradation constants. Many attempts have been made to enhance the biodegradation of these persistent pharmaceuticals in water bodies through enzymes.[17] The immobilization of enzymes on ACNPs can retain their original activities for an extended time compared to free-form enzymes and possess high recovery and recyclability properties.[18] Al-sareji *et al.*[19] successfully produced loaded ACNPs by laccase oxidoreductase enzyme from pomegranate peels to remove a range of pharmaceutical water pollutants including diclofenac, amoxicillin, carbamazepine, and ciprofloxacin using

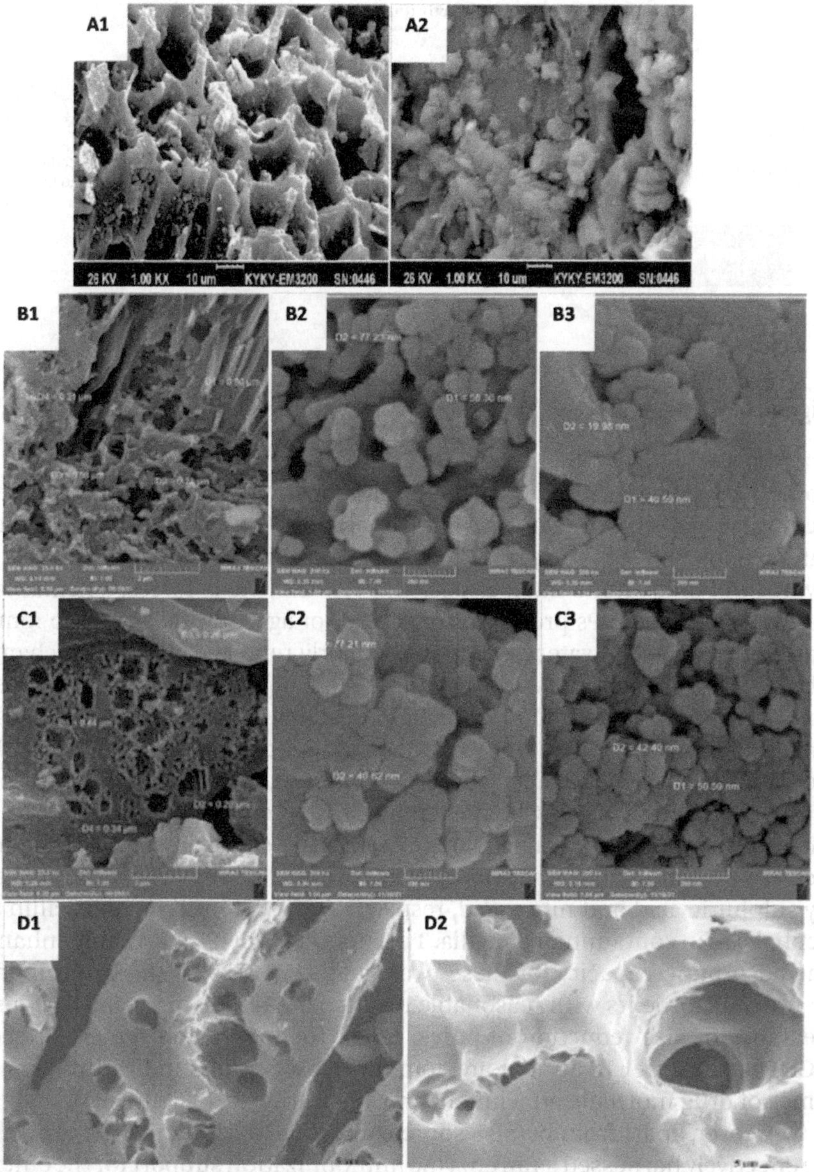

Figure 6.4 SEM images of: (A) ACNPs from *Azolla filiculoides*; (A1) before and (A2) after adsorption of ampicillin. Reproduced from ref. 15 with permission from Springer Nature, Copyright 2021. (B) ACNPs from aloe vera treated with sulphuric acid; (B1) before and after adsorption of (B2) amoxicillin and (B3) cephalexin; (C) ACNPs from aloe vera treated with nitric acid (C1) before and after adsorption of (C2) amoxicillin and (C3) cephalexin. Reproduced from ref. 16 with permission from Elsevier, Copyright 2022. (D) ACNPs from pomegranate peel; (D1) before and (D2) after immobilization with laccase enzyme. Reproduced from ref. 19, https://doi.org/10.1038/s41598-023-38821-3, under the terms of the CC BY 4.0 license http://creativecommons.org/licenses/by/4.0/.

synergistic adsorption and enzymatic degradation. The laccase immobilization reduced the surface area for adsorption but added more active sites for the degradation of drugs (see Figure 6.4).[19] Additional incorporation of the biodegradation method in the ACNPs common adsorptive mechanism offers better economic benefit owing to their stability and repeatability for pharmaceutical pollutants.

The efficiency of ACNP nanoadsorbents for emerging herbicide water pollutants were also explored. Chlorinated phenoxy acetic acid or specifically 2,4-dichlorophenoxyacetic acid (2,4-DPA) is one of the most widely used herbicides to combat broad-leaved weeds and pests in various types of crops.[20] It is highly soluble in water due to its polar properties and the presence of an acidic carboxylic acid group. Despite the many conventional adsorbents that have been examined, the ACNP has shown the highest adsorption capacity for 2,4-DPA herbicide removal (Jayaraman *et al.*, 2018).[21] Different types of agricultural wastes such as orange peel,[22] orange pulp,[23] groundnut shell,[24] and canola stalk[25] have been used for the removal of 2,4-DPA herbicide from aqueous medium. Rambabu *et al.*[26] synthesized ACNPs using date palm coir waste activated by potassium hydroxide with a maximum adsorption capacity of 50.25 mg g^{-1} for 2,4-DPA removal superior to other activated carbon adsorbents. It was postulated that the adsorption/desorption mechanism of 2,4-DPA on the ACNPs surface was heavily influenced by pH. In an acidic environment, the surface of ACNPs will be positively charged, thus inducing electrostatic interactions with neutral 2,4-DPA in addition to common adsorptive interactions such as π–π interactions, hydrogen bonding, and van der Waals forces. In a basic environment, the 2,4-DPA molecule will be deprotonated and the surface of ACNPs will be in negative charge inducing electrostatic repulsion *i.e.* desorption.[26]

Magnetization of ACNPs is one of the strategies used to facilitate separation of adsorbent of treatment from the liquid matrix. This is a very helpful method to increase adsorbent recovery and regeneration as well as controlling the adsorption process. de Freitas *et al.*[27] modified the ACNPs with impregnated magnetite Fe_3O_4 *via* a co-precipitation method for the removal of thiamethoxam (TMX) pesticide from water bodies.[27] Figure 6.5 illustrated the magnetization strength of powdered ACNPs and its suspension in water. The recovery process of the magnetic adsorbents is usually by imposition of a magnetic field. Kumar *et al.*[28] prepared a magnetic acid ACNP from the shells of *Sterculia villosa Roxb* for adsorptive removal of ciprofloxin. Figure 6.6 depicts the surface morphology of magnetic acid ACNPs with aggregates of iron oxides formed by co-precipitation of ferrous (Fe^{2+}) and ferric (Fe^{3+}) ions. The surface became smoother with less irregularities and cavities for post adsorption of ciprofloxacin.[28] Sharma *et al.*[29] reported the preparation of magnetic nanocomposite material from algal-mediated ACNPs and iron nanorods for the removal of diclofenac sodium, a non-steroidal anti-inflammatory drug (NSAID) with maximum adsorption efficiency of 82.1%. The sorption capacity decreased only 8.2% after five cycles of regeneration and reuse compared to the original adsorption.[29]

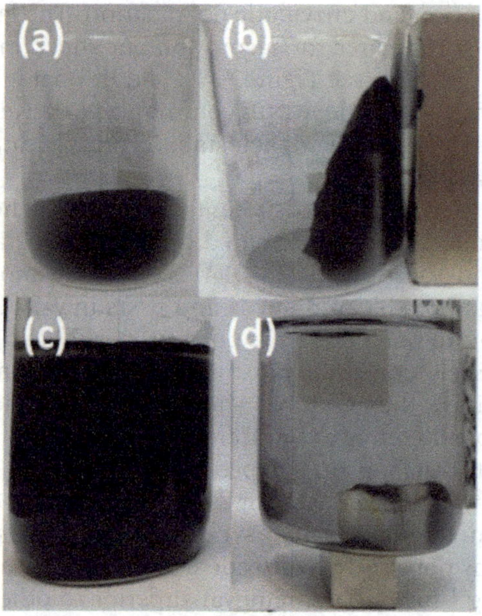

Figure 6.5 Illustration of (a) powdered magnetic ACNPs-Fe_3O_4 nanocomposite and (b) its magnetization effect; (c) suspension of magnetic ACNPs-Fe_3O_4 nanocomposite in TMX contaminated water; and (d) its magnetization effect.[27] Reproduced from ref. 27 with permission from Springer Nature, Copyright 2022.

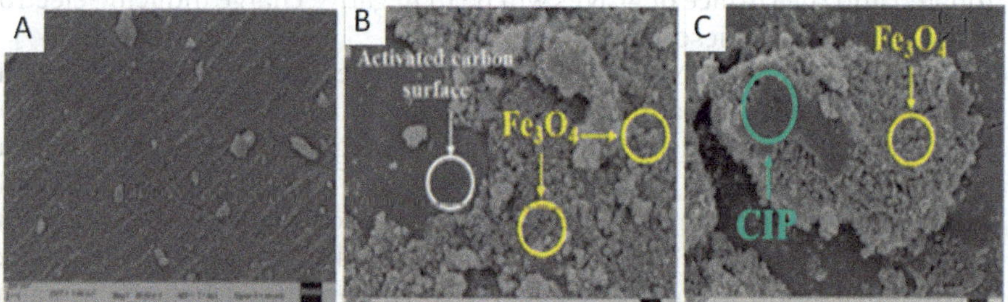

Figure 6.6 FESEM images of (A) ACNPs from the shells of *Sterculia villosa Roxb*; (B) co-precipitation of magnetite Fe_3O_4 on the ACNPs and (C) the adsorption of ciprofloxacin.[28] Reproduced from ref. 28 with permission from Elsevier, Copyright 2022.

6.2.2 Nanodiamonds

Diamond represents a metastable carbon allotrope, characterized by a face-centred cubic crystal structure.[30] It has exceptional physical properties including unrivalled hardness and thermal conductivity amongst bulk materials. Nanodiamonds, which are diamonds with nanoscale dimensions, were first discovered by Soviet scientists in the 1960s. These nanodiamonds were found in soot generated by detonating a 60% oxygen-deficient trinitrotoluene (TNT)/40% hexogen composition in an inert medium, without any additional carbon source.[31,32] Structurally, nanodiamonds consist of a diamond core encompassed by an amorphous carbon shell. Through X-ray diffraction and small-angle X-ray scattering, the average particle size of nanodiamonds was determined to be 4–5 nm, allowing them to form colloidal suspensions of individual diamond particles.[31]

Various synthesis methods can produce nanodiamonds, including impact events such as detonation or meteoritic impacts, high-energy ball milling of high-pressure high-temperature diamond microcrystals, laser ablation, plasma-assisted chemical vapor deposition (CVD), ion irradiation of graphite, and ultrasound cavitation. Among these, detonation-synthesized nanodiamonds have already been commercialized and find frequent applications in polymer composites, energy, environmental, and electrical fields.[31]

Boron-doped diamond (BDD) electrodes can be categorized into the following classes: (1) ultra-nanocrystalline (UCN) with a grain size of <10 nm, (2) nanocrystalline (NC) with a grain size in the range 10 nm to 1 μm, and (3) microcrystalline (MC) with a grain size of >1 μm.[33] The grain size can be controlled from the film thickness.[33] As the film thickness increases, the grain size also increases as depicted in the illustration published in the original article.

6.2.2.1 *Nanodiamonds for Water and Wastewater Treatment*

The current demand for advanced water treatment technology with higher oxidizing power has intensified due to the challenges posed by processing non-biodegradable organic compounds generated by industrial processes when using existing methods. Electrochemical water treatment offers a cost-effective, safe, and environmentally friendly approach. Specifically, electrolytic oxidation is a widely employed method for decomposing pollutants utilizing an inert electrode without the need for additional reactants. In this process, the choice of anode material is crucial, as it plays a pivotal role. Therefore, materials with excellent electrical power efficiency and stability are sought after. Among the commonly used materials for treating challenging bio-refractory wastewater over the past decades, the boron-doped diamond (BDD) anode stands out as a prominent choice.[31] Considering environmental sustainability, it is noteworthy that BDD electrodes consist solely of elements abundant on Earth, namely carbon, boron, and oxygen.[34]

Numerous investigations have been conducted to explore the potential of BDD as anode materials. These studies have demonstrated the extensive utilization of BDD as anodes, primarily attributed to their remarkable attributes, including high oxidation capacity, excellent current efficiency, chemical and electrochemical stability, prolonged operational lifespan, corrosion and mechanical resistance, thermal conductivity, low background current, long service life, wide potential window, and minimal electrode fouling.[31,34,35]

The boron concentration usually plays an important role in the electrical properties and performance of BDD electrodes. The precise boron concentration significantly influences the conductivity, charge transfer efficiency, and overall electrochemical behaviour of the electrode. Properly adjusting the boron content is vital to achieve optimal electrical conductivity and enhance the electrode's electrochemical performance. Gonzaga and colleagues[36] investigated the potential of BDD electrodes in removing penicillin G, total organic carbon (TOC), and chemical oxygen demand (COD) from urine matrix which could have been discharged into hospital wastewater. They reported that BDD electrodes with low boron concentrations which were 100 and 200 ppm exhibited a higher oxidation efficiency than electrodes with high boron doping where they have achieved almost 100% removal of the antibiotic. A similar conclusion was made for the TOC and COD. Low-doped diamond electrodes have a high diamond (sp^3)/graphite (sp^2) ratio which could have influenced the removal efficiency of the pollutants. The electrical conductivity of the material is influenced by various factors, including lattice hydrogen, sp^2-carbon impurities, and dangling bonds, in addition to the doping level. Concerning non-diamond sp^2 carbon, an increase in boron content within the film leads to a reduction in the size of crystallites, resulting in the appearance of impurity phases. Consequently, this impacts the electron-transfer kinetics of redox systems differently, as well as the background current and potential window. In general, the higher the sp^2 content, the smaller the effective working potential window and the larger the background current.

The main oxidizing species that facilitates the removal of pollutants is the hydroxyl radicals generated on the surface of the BDD electrodes. It then creates an arsenal of oxidizing agents produced on BDD electrodes during the anodic oxidation during water or wastewater electrochemical treatment processes which has been summarized in Table 6.1.[35]

These products may oxidize the existing pollutants in water and wastewaters due to their oxidizing capabilities. In order to easily understand the processes that are occurring on the surface of BDD electrodes during anode oxidation, Figure 6.7 illustrates all possible reactions that may facilitate the elimination of the pollutants.

To reduce the cost, stainless steel is often employed as the cathode. However, BDD electrodes can also be used as a cathode in order to enhance the removal efficiency of the water contaminants. Oxygen (O_2) reduction usually occurs at the cathode to produce hydrogen peroxide (H_2O_2). Adding a catalytic amount of Fe^{2+} to the contaminated water to react with the generated H_2O_2 will form Fe^{3+} and $^{\bullet}OH$ in the bulk from the Fenton reaction under acidic conditions.[35]

Nevertheless, a significant limitation lies in the high manufacturing costs per unit surface area associated with BDD electrodes when produced using chemical vapor deposition or similar surface deposition methods.[34] Therefore, BDD nanoparticles have been proposed as an alternative manufacturing process that reduces the costs by more than 1000-fold.[34] Figure 6.8 shows

Table 6.1 The processes that may occur on the surface of BDD electrodes during anode oxidation for the treatment of water and wastewater.[35] Adapted from ref. 35 with permission from Elsevier, Copyright 2021[a]

Initial process:	
$BDD + H_2O \rightarrow BDD(\bullet OH) + e^- + H^+$	2.8 V/SHE
Subsequent potential processes from the initial process:	
$HSO_4^- + BDD(\bullet OH) \rightarrow SO_4^- + OH^-$	
$SO_4^{2-} + BDD(\bullet OH) \rightarrow SO_4^- + OH^-$	
$CO_3^{2-} + BDD(\bullet OH) \rightarrow CO_3^- + OH^-$	
$PO_4^{3-} + BDD(\bullet OH) \rightarrow PO_4^{2-} + OH^-$	
Subsequent potential processes from the radicals:	
$SO_4^{\bullet -} + SO_4^{\bullet -} \rightarrow S_2O_8^{2-}$	2.01 V/SHE
$CO_3^{\bullet -} + CO_3^{\bullet -} \rightarrow C_2O_6^{2-}$	1.80 V/SHE
$PO_4^{\bullet 2-} + PO_4^{\bullet 2-} \rightarrow P_2O_8^{4-}$	2.07 V/SHE
When sodium chloride is used as a supporting electrolyte:	
$2Cl^- \rightarrow Cl^2 + 2e^-$	
$Cl^- \rightarrow Cl_{ads}^{\bullet} + e^-$	
$Cl_{ads}^{\bullet} + Cl^- \rightarrow Cl_2 + e^-$	
$Cl^- + BDD(\bullet OH) \rightarrow Cl^{\bullet} + OH^-$	
$Cl^- + BDD(\bullet OH) \rightarrow ClO^- + 2H^+ + 1e^-$	
$ClO^- + BDD(\bullet OH) \rightarrow ClO_2^- + H^+ + 1e^-$	
$ClO_2^- + BDD(\bullet OH) \rightarrow ClO_3^- + H^+ + 1e^-$	
$ClO_3^- + BDD(\bullet OH) \rightarrow ClO_4^- + H^+ + 1e^-$	

[a]SHE = standard hydrogen electrode; H_2O = water; BDD(•OH) = hydroxyl radical(s) on BDD electrodes; H^+ = proton(s); e = electron(s); HSO_4^- = monoacid sulfate; $SO_4^{\bullet -}$ = sulfate radical ions; OH^- = hydroxide ions; SO_4^{2-} = sulfate ions; CO_3^{2-} = carbonate ions; PO_4^{3-} = phosphate ions; $S_2O_8^{2-}$ = persulfate ions; $C_2O_6^{2-}$ = percarbonate ions; $P_2O_8^{4-}$ = perphosphate ions; Cl^- = chloride ions; Cl_2 = chlorine; ClO^- = hypochlorite ions; ClO_2^- = chlorite ions; ClO_3^- = chlorate ions; and ClO_4^- = perchlorate ions.

the scanning electron microscopy (SEM) images of nano BDD electrodes at different magnifications. The nano BDD electrodes were fabricated using an ink casting method to ensure uniform coatings where the precursor ink solutions were prepared using different ratios of the aqueous suspension of nano-BDD in Nafion® and isopropanol.[34]

Traditional water treatment technologies often prove ineffective in removing persistent and recalcitrant emerging pollutants, leading to their accumulation in water bodies and potential long-term adverse effects. To address this pressing issue, researchers and environmental scientists have turned their attention to advanced and innovative technologies, among which BDD electrodes have emerged as a promising solution.

In recent years, extensive research has been conducted to investigate the feasibility of using BDD electrodes for the removal of emerging contaminants from water sources. The electrochemical oxidation process facilitated

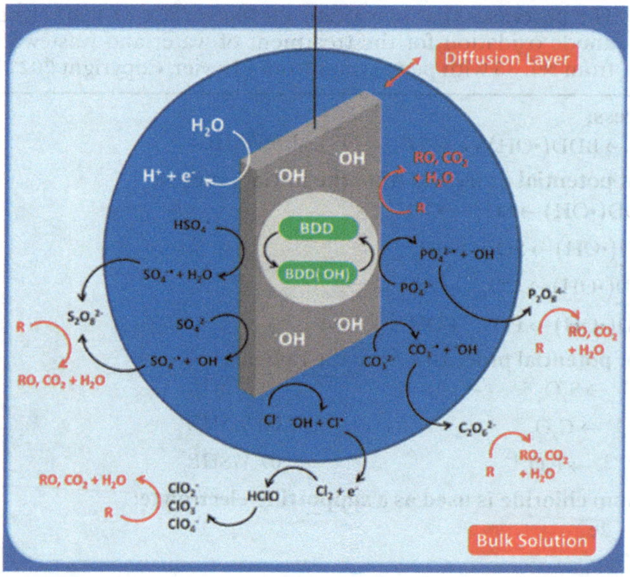

Figure 6.7 Electrochemical processes that are occurring on the surface of a BDD electrode during an anodic oxidation to treat water or wastewater from any pollutants which are denoted by R. CO_2 = carbon dioxide and RO = oxidized pollutants.[35] Reproduced from ref. 35 with permission from Elsevier, Copyright 2021.

Figure 6.8 SEM images of (A) a macroscopic BDD electrode film, (B and C) nano BDD electrodes at 2500× and 10 000× magnification respectively. Reproduced from ref. 34 with permission from Elsevier, Copyright 2022.

by BDD electrodes has shown remarkable efficiency in degrading various organic pollutants, transforming them into harmless by-products. Additionally, BDD electrodes' ability to generate hydroxyl radicals and other powerful oxidants during the electrochemical process further enhances their contaminant removal capabilities. As a result, BDD-based electrochemical treatment systems hold great promise as a sustainable and environmentally friendly technology to combat the growing issue of emerging contaminants in water bodies.

A study reported the use of Macroscopic BDD thin films are characterized by rough surfaces, resulting from the growth of diamond polycrystals within the vacuum chamber of CVD. Figure 6.8(A) presents a SEM image of a macroscopic BDD electrode, revealing the faceted crystalline structure of BDD. In contrast, the as-prepared nano-BDD electrodes with a lower concentration of 0.3 wt% boron exhibits notable distinctions. When compared to smooth Si wafer substrates, the nano-BDD electrode displays a uniform distribution of nanoparticles on the Si substrate, contributing to increased roughness and a higher interfacial surface area, as quantified using optical profilometry. The nanoparticles in the nano-BDD electrode possess a spherical morphology, with an average size of 50–80 nm, and their homogeneous distribution on the substrate surface remains consistent. The most significant observation with the nano-BDD electrodes is the considerable increase in the electroactive area surface due to the nanostructured morphology of the coatings. Having a high surface area could lead to a higher efficiency in removing water and wastewater contaminants.

The efficiency of phenol removal using the nanocrystalline BDD thin film produced on cemented carbide (WC-Co) was evaluated in replicated experiments. The average COD reduced from the initial concentration of 5795 mg L^{-1} to 85 mg L^{-1}, indicating a significant reduction in phenol concentration. The average current efficiency for phenol oxidation was approximately 46%, suggesting a good mineralization capacity of the nanocrystalline BDD/WC-Co electrode for phenol with a high concentration. These results demonstrate the effectiveness of the electrode in removing phenol from wastewater.[37] Phenol and landfills leachate seem to be commonly used as analytes amongst articles that use BDD electrodes for electrochemical removal of water pollutants.[38]

Several studies have explored the use of BDD electrodes to eliminate emerging pollutants from water samples and model systems (see Table 6.2). An overview of these research efforts was discussed by Oliveira *et al.*[39] Nevertheless, there is still room for further work to enhance the efficiency of pollutant removal and reduce the production costs associated with BDD electrodes.

In conclusion, nanodiamond has great potential in treating wastewater because of its efficiency in removing emerging pollutants in water and wastewater because of the radical ions that are produced on the surface of the electrodes. However, there is still much work that needs to be done to find ways to reduce the cost of producing doped diamond electrodes and to further increase the removal efficiency of the pollutants.

6.2.3 Fullerenes

In 1985, Kroto *et al.* discovered fullerene C60,[40] which belongs to the fullerene family and is recognized as one of its most prevalent and stable members.[41] These carbon nanostructured materials are known for their small size, stability, and exceptional properties, including strong antioxidant capacity and high conductivity.[42] With a large surface-to-volume ratio and electron mobility that allows them to accept more electrons at room temperature, fullerene C60 demonstrates appealing performances like graphene and carbon

Table 6.2 Pollutant removal using BDD electrodes along with their operation conditions and methodology of removing the pollutants. This table also summarizes the pollutant removal percentage. Adapted from ref. 35 with permission from Elsevier, Copyright 2022, and ref. 39 with permission from Elsevier, Copyright 2022.

Anode material	Pollutant	Methodology	Operating conditions	Real sample	Pollutant removal percentage
BDDE	Bisphenol A	Electrolysis	20 mg L^{-1} | 300, 500, 700, and 1000 A m^{-2} | 22–25 °C	Simulated and real sanitary landfill leachate	77% in chloride medium and 86% in sulfate medium
BDDE	Bisphenol A	Electrolysis	50 mg L^{-1} | 9.04 mA cm^{-2} | flow rate = 400 mL min	$NaNO_2$	From 8.2% up to 63.4% of TOC
Fe and BDDE	Bisphenol A diisooctyl phthalate	PEF	H_2O_2 from 0 to 600 mg L^{-1} | 0.9 A |flow rate of 0.6–0.8 min	Wastewater from landfill leachate	77.9% of COD, 71.5% of TOC, and 76.3% radiation absorbance
BDDE and MMO (TV RuO_2–IrO_2)	Bisphenol A	Electrolysis	5 µmol L^{-1} bisphenol A | 30 mmol L^{-1} $NaClO_4$ or Na_2SO_4 or NaCl and 1 mmol L^{-1} $Na_2S_2O_8$ | 5 mA cm^{-2}	Wastewater treatment plant and anaerobic–aerobic process of landfill leachate	85% for BBD and 60% for DSA
BDDE	*p*-Methoxyphenol | *p*-nitrophenol | *p*-chlorophenol | *p*-chlorophenol	Electrolysis	0.2 mol L^{-1} $NaSO_4$ | 20 mA cm^{-2}	—	100% for *p*-chlorophenol | 73.6% for *p*-methoxyphenol
BDDE	Bisphenol A	Electrolysis	5 mmol L^{-1} bisphenol A in NaCl or in Na_2SO_4 solutions | 42.7 mA cm^{-2} | flow rate = 384 mL min^{-1} | pH = 7 or 10 | 6 and 20 °C	—	90%
BDDE	Bisphenol A | Bisphenol B | Bisphenol F | Bisphenol S	UVC-assisted electrolysis and UV-photolysis	10 mg of pollutants in 0.2 mol L^{-1} $NaSO_4$ | UVC irradiance of 1.425 ± 0.064 mW cm^{-2} | 10, 20, and 30 mA cm^{-2}		About 100%

BDDE	Bisphenol A | Bisphenol AF | Bisphenol S	Solar-assisted electrolysis	5 µmol L^{-1} of compounds | 5 mA cm^{-2} | 25 °C	Wastewater from a wastewater treatment plant	64% for bisphenol AF, 76% for bisphenol A, and 52% for bisphenol S
BDDE	Methyl paraben (MeP)	Electrolysis and photoelectrolysis	100 MeP | 3 g L^{-1} Na_2SO_4 | 3.7 g L^{-1} NaCl electrolysis | 15–60 mA cm^{-2} | Hg vapor lamp | $\lambda = 254$ nm | 25 °C	—	100% for irradiated and non-irradiated electrolysis, 17% for chloride medium, 5% for sulfate medium, and 100% for electrolysis and photoelectrolysis
BDDE	Methyl paraben	AO-H_2O_2, AO-H_2O_2/UVC, EF, and PEF	0.50 mmol L^{-1} MeP + 50 mmol L^{-1} Na_2SO_4 | pH 2. 33.3 mA cm^{-2} | 0.1, 0.25, and 0.50 mmol L^{-1} of Fe^{2+} | 5 W UVC lamp (19 W m^{-2})		26.1% for AO-H_2O_2, 38.4% for AO-H_2O_2/UVC, 45.9% for EF, and 90.1% for PEF
BDDE stainless steel platinum	Butyl paraben	Electrolysis	250–2000 µg L^{-1} or 0.1 mol L^{-1} Na_2SO_4 | 1.4–143 mA cm^{-2}	Secondary treated wastewater, bottled water, surface water. Ultrapure water, and ultrapure water spiked with humic acid	
BDDE	Potassium hydrogen phthalate (KHP)	Electrolysis	1 mmol L KHP in 0.1 mol L^{-1} Na_2SO_4		42%
BDDE	Dibutyl phthalate (DBP)	Electrolysis	5 mg L^{-1} DBP | 0.8 mol L^{-1} Na_2SO_4 | 30 mA cm^{-2} | 25 °C		91%–34.5% using TBA | 10% using MeOH

(*continued*)

Table 6.2 (*Continued*)

Anode material	Pollutant	Methodology	Operating conditions	Real sample	Pollutant removal percentage
BDDE	Carbamazepine, diclofenac, trimethoprim, and venlafaxine	Electrolysis	21.5 ± 1 °C | pH 7.2	Effluent wastewater obtained from model of one equivalent domestic wastewater treatment plant	An increase from 10 000 ppm to 15 000 ppm caused a decrease of removal efficiency
BDDE	Reactive dye rubin F-2B	Electrolysis	NaCl and $NaSO_4$		Higher amounts of dopant on the surface of BDD resulted in the higher efficiency of dye removal in NaCl and Na_2SO_{44}
BDDE	1-Butyl-3-methylimidazolium chloride (BMImCI)	Electrolysis	0.3 L and 0.25 g L^{-1} of BMImCl in 0.1 mol L^{-1} of Na_2SO_4		A higher sp3/sp2 ratio results in a more efficient process in the degradation and secondary compounds production
BDDE	Rhodamine B (RhB)	Electrolysis	1.0 L and 2.09 × 10 4 mol L^{-1} of RhB, Na_2SO_4 | 97 mA cm^{-2}		The higher the content in diamond/graphite the greater are the TOC and COD decay rates favoring electrochemical combustion of RhB
BDDE	Phenol	Electrolysis	0.04 M phenol and sulfuric acid | 100 mA cm^{-2}		63 of COD removal after 240 min of electrolysis
BDDE	*m*-Cresol	Electrolysis	Na_2SO_4		>70% of *m*-cresol removal and 45% of TOC elimination
BDDE	Dimethyl sulfoxide (DMSO)	Electrolysis	0.1, 0.2, or 0.3 M NaOH | 0.1 A cm^{-2} | 5 hours		Almost 100% of TOC removal after 5 hours of electrolysis

BDDE	Brilliant green dye	Electrolysis	10 mA cm^{-2} | 0.1 mol L^{-1} K_2SO_4 | 25 °C.		More than a 90% of color removal and a 100% of degradation with the lowest current density applied
BDDE	Reactive Blue 19	Electrolysis	Strong acid or strong alkaline solutions | 70 °C		100% of color removal and 55% of mineralization in NaCl, used as supporting electrolyte
BDDE	Glucose	Electrolysis	0.2 M Na_2SO_4 | 30 mA cm^{-2}		~60% of COD removal after 24 h of electrolysis
BDDE	18 Wastewaters treatment plants (WWTPs)	Electrolysis	Constant applied voltage = 30 V | 20 mA cm^{-2}.	Wastewater in Slovak and Czech cities	>51% reduction in the concentration of 14 drugs present in the WWTPs
BDDE	Methyl red	Electrolysis UV radiation assisted	5 mA cm^{-2} in 0.051 M NaCl	—	70% removal. Can be increased with the coupling of UV radiation

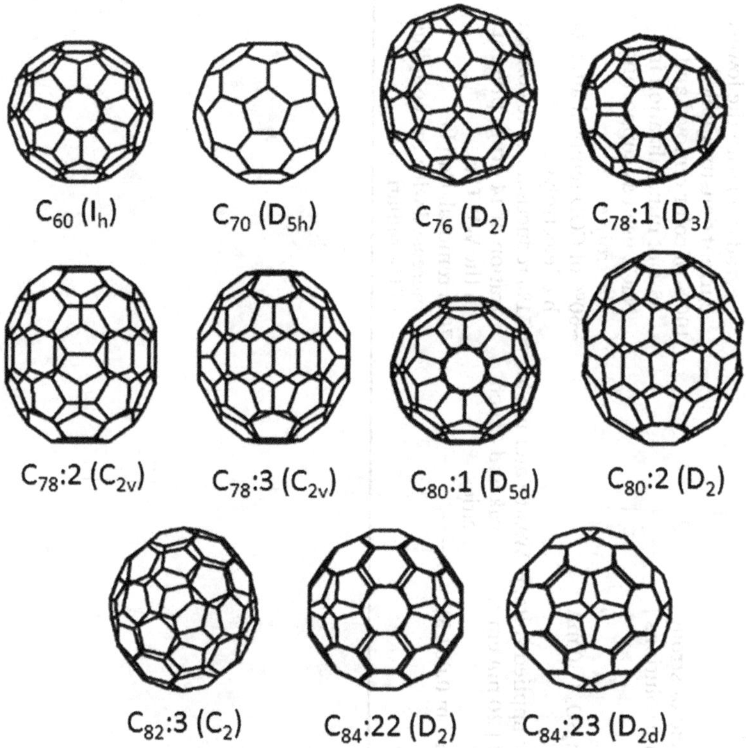

Figure 6.9 Range of fullerene derivatives. Reproduced from ref. 43 with permission from Springer Nature, Copyright 2015.

nanotubes, particularly in water treatment applications. Researchers have also synthesized a range of fullerene C60 derivatives, such as C540, C70, C76, C82 and C84 (see Figure 6.9),[43] to explore their full potential and their possible uses in addressing emerging water pollution challenges. Among these forms, the C60 and C70 were more extensively investigated than others due to their properties that are suitable for various applications.

Researchers have revealed various experimental approaches to synthesize fullerene, including laser vaporization of carbon, electric arc heating of graphite, resistive arc heating of graphite, laser irradiation of polycyclic hydrocarbons, and several others.[44] Subsequently, substantial amounts of fullerene, particularly C60, have been produced using methods like arc heating of graphite and laser irradiation of polyaromatic hydrocarbons. The exceptional properties of fullerene have drawn considerable attention for wastewater treatment and purification, whether in the form of derivatives, composites, or modified versions. Numerous reports have explored the design and performance of fullerene-based desalination, ultra- and nanofiltration, pervaporation, and many more. Fullerenes are often employed as adsorbents in industrial wastewater treatment due to their large surface area, which aids in the removal of heavy metal ions which are known contaminants.[45] Meanwhile fullerene derivatives with hydrophobic properties, combined with photocatalysts, offer a means to address pathogens. However, there remains a lack of research on optimizing the utilization of fullerenes for removing emerging water pollutants.

6.2.3.1 Fullerenes for Emerging Water Pollutant Removal

The utilization of fullerenes as electrodes has been explored for the detection of anti-inflammatory drugs, specifically diclofenac (DF), naproxen (NX), and ibuprofen, all of which are considered to be emerging water pollutants. A study conducted by Motoc *et al.*[46] highlights the potential of utilizing fullerene carbon nanofiber electrodes for detecting these anti-inflammatory drugs in the surface water of the Bega river in Timisoara, Romania.

The fullerene carbon nanofiber electrodes exhibited favorable qualities such as reproducibility, stability, and the capability to simultaneously detect diclofenac, naproxen, and ibuprofen in aqueous solutions. This was achieved through various detection techniques including differential pulsed voltammetry, chronoamperometry, and multiple-pulsed amperometry.[46] This electrode proved to be well-suited to the creation of a detection methodology based on simultaneous differential pulsed voltammetry. This approach successfully achieved remarkably low detection limits of 0.230 nM for diclorenac, 0.310 nM for naproxen, and 0.180 nM for ibuprofen.[46]

In a separate investigation conducted by Elessawy *et al.*,[47] a functionalized magnetic fullerene nanocomposite was synthesized using a one-pot method. This synthesis involved the catalytic thermal decomposition of eco-friendly poly(ethylene terephthalate) bottle waste as the raw material, with ferrocene serving as both a catalyst and a precursor for magnetite (see Figure 6.10). Subsequently, the efficacy of these magnetic fullerene nanocomposites was assessed in terms of their ability to adsorb Ciprofloxacin antibiotic.

The ionizable functional groups located on the surface of the magnetic fullerene that was synthesized had the capability to engage with and attach to the Ciprofloxacin molecules. This interaction led to the creation of structures characterized by multiple carbon–carbon bonds, achieved by displacing the original oxygen and hydrogen atoms present (see Figure 6.11). The magnetic fullerenes that were prepared demonstrated a substantial rate of adsorption for the Ciprofloxacin antibiotic from aqueous solutions. Additionally, the adsorption kinetics were excellently described by the pseudo-second-order kinetic model. The equilibrium data gathered at varying temperatures displayed an excellent fit with the Langmuir and Freundlich isotherm models. Upon statistical analysis, it was deduced that the process of adsorbing Ciprofloxacin was spontaneous, exothermic, and displayed a degree of randomness.

Given the considerable potential of fullerene, particularly C60, in aiding the mitigation of emerging water pollutants, it is crucial to comprehensively investigate the potential toxic effects of C60 in freshwater ecosystems.

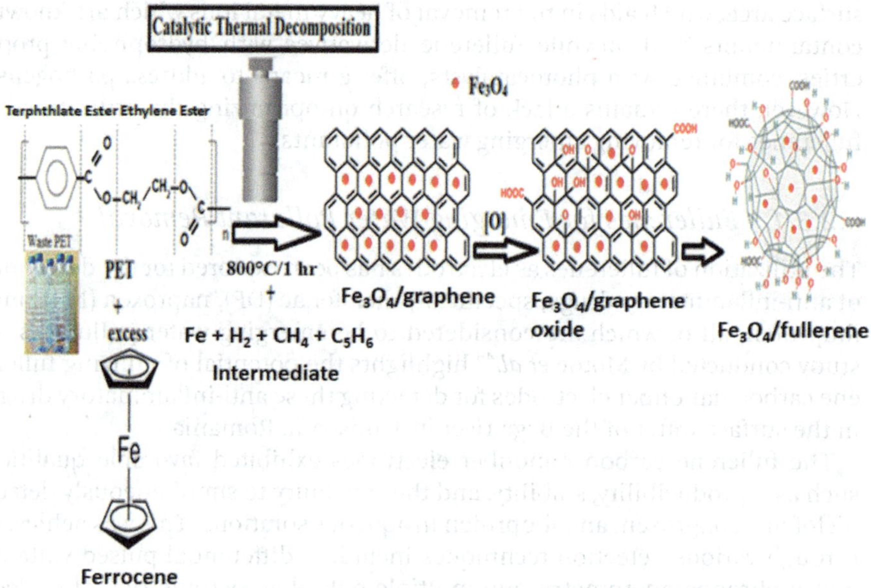

Figure 6.10 Schematic illustration for the one-step formation of functionalized magnetic fullerene nanocomposites. Reproduced from ref. 47 with permission from Elsevier, Copyright 2020.

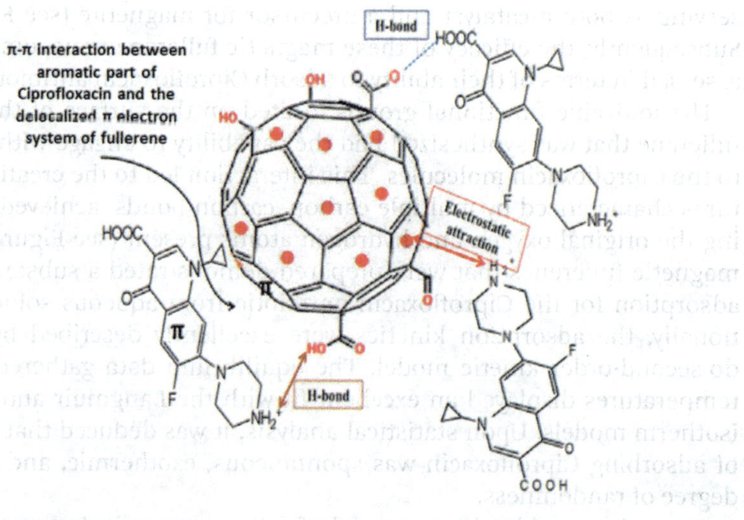

Figure 6.11 Different types of interactions involved in the adsorption of ciprofloxacin on magnetic fullerene. Reproduced from ref. 47 with permission from Elsevier, Copyright 2020.

To address this concern, Sumi and Chitra[48] conducted a study that aimed to assess the impact of C60 exposure on the ecosystem of freshwater fish, specifically *Anabas testudineus*.

Exposure to fullerene C60 was found to induce significant alterations in the behavior of the fish. These alterations included changes in swimming patterns, disruptions in schooling behavior, instances of air intake and surfacing, in addition to noticeable morphological changes. Moreover, hematological changes were observed, including a significant substantial reduction ($P < 0.05$) in both erythrocyte and leukocyte counts, as well as a decrease in hemoglobin concentration.[48] These findings indicate that sublethal concentrations of fullerene C60 exert a toxic effect on the fish species *A. testudineus* by disrupting its normal physiological processes. Consequently, the presence of this carbon molecule in the environment could potentially impact the overall health of the ecosystem.

6.2.4 Graphene and Graphene Oxides

The research focusing on the utilization of graphene materials including pristine graphene, graphene oxides, functionalized graphene and graphene-based composites has been aggressively reported in the scientific communities for various applications including for water remediation owing to the urgent need of technological advancements in tackling a wide range of novel or emerging pollutants in water systems. Graphene materials including graphene and graphene oxides have garnered substantial attention targeted as potential and efficient materials for water purification due to their large surface area, tunable surface chemistry, and unique interactions with a range of contaminants. These intriguing attributes have propelled graphene-based materials into the spotlight as promising candidates for the removal of emerging water pollutants. Although graphene ideally consists of pure sp^2 carbon bonds, graphene practically varies according to thickness, size, and surface defects depending on the fabrication process which in turn governs the types of graphene such as pristine graphene, graphene flakes, graphene oxides, or reduced graphene oxides.

Many scientists elucidated that the potential interaction mechanism possessed by graphene-based nanomaterials such as electrostatic interactions, π–π stacking, and surface adsorption has collectively facilitated the adsorption and immobilization of a wide spectrum of pollutants including the novel or emerging contaminants. Therefore, many scientists have endeavoured to modify the surface properties of graphene with the objective of enhancing its efficiency and facilitating its reusability.[49] Various types of emerging pollutants have been investigated and, in most cases, reduced graphene oxide and pristine graphene exhibited comparatively lower adsorption capacities when compared to functionalized graphene or graphene oxide. This phenomenon might be attributed to the increased hydrophobicity of the surface, characterized by a reduction in the number of oxygen functional groups. Consequently, the adherence of emerging pollutants in aqueous environments is impeded. However, due to the presence of its delocalized electrons and its extensive surface area, graphene also demonstrates suitability for the elimination of organic compounds that consist of benzene rings and π–π stacking. For instance, the use of graphene oxide has been employed to remove tetracycline, which has four aromatic rings, each of which is characterized by a unique functional group, namely phenol, aldehyde, ketone, and amino. The adherence of the material to the graphene oxide surface is facilitated by cation bonding and π–π interaction.[50] One of the challenges associated with graphene pertains to its hydrophobic nature, which poses difficulties in terms of its recyclability. Graphene exhibits limited efficacy in adsorbing polar components that possess hydrophilic functions. Butler and coworkers[51] investigated the potential of nanostructured porous graphene for the removal of various pharmaceutical emerging pollutants from water including ciprofloxacin (CIP), medicines atenolol (ATL), gemfibrozil (GEM), carbamazepine (CBZ), diclofenac (DCF), and ibuprofen (IBP). They revealed that the nanostructured graphene able to provide high clearance efficiencies for these emerging pharmaceutical pollutants even at low graphene dosage.[51]

Recently, Kaur *et al.*[52] reported the synthesis of a reduced graphene oxide–tungsten trioxide (rGO–WO_3) heterostructure as an efficient adsorbent for the removal of fluoroquinolone antibiotic, namely levofloxacin (LVX), from wastewater. They revealed the significance of electrostatic interactions for the adsorption of LVX onto the rGO–WO_3 heterostructure. The deposition of WO_3 on the surface of rGO prominently avoided the restacking of rGO nanosheets which further formed a strong interfacial contact between WO_3 nanorods and rGO. These thus allow the improved performance of the adsorbent for the removal of the pollutant due to larger surface area and offer more reactive sites resulting from reduced stacking or agglomeration. Roberts *et al.*[53] investigated the utilization of graphene nanoplatelets as a rational interface for the removal of cyanotoxins, namely microcystin-LR (MC-LR). The interaction between the sp^2 hybridized carbon of graphene and aromatic ring of MC-LR played a significant role in the adsorption mechanism. In addition, the author also revealed that the results of the fitting of the intra-particle diffusion model suggests that the graphene material possessing high mesoporosity exhibits significant improvements in both adsorption capacities and kinetics. Furthermore, the use of density functional theory (DFT) computations demonstrated a favorable alignment between the hypothesized π-based interactions and the experimental findings obtained by nuclear magnetic resonance (NMR) techniques. These experimental methods were employed to investigate the principal interactions occurring between graphene and the MC-LR adduct. The findings of this work provide evidence supporting the prominent role of π-interactions between the aromatic ring on MC-LR and graphene sp^2 orbitals.[53] The removal of emerging pollutants such as tetracycline utilizing graphene derived from liquid phase exfoliation technique was reported by Sankar *et al.*[54] *via* selective ion transport graphene membranes. Briefly, the authors use high-shear mixing of pristine graphene flakes in cyrene as a method for exfoliating graphene flakes. This approach offers an alternative to the use of common solvents in liquid-phase exfoliation, including *N*-methyl-2-pyrrolidone, dimethylformamide, and dimethyl sulfoxide. These solvents are known to possess inherent safety concerns, toxicity, and potential dangers to both human health and the environment. Therefore, the technique introduced by the author might pave the way towards a green strategy of graphene production using a liquid phase exfoliation technique for emerging contaminant removal. The dispersion of graphene flakes in cyrene is subjected to vacuum filtration at varying volumes in order to produce membranes of increasing thicknesses. These membranes possess a homogeneous surface composed of tightly packed graphene flakes, exhibiting no pinholes. In addition, the membranes exhibit structural integrity, with a negligible increase in thickness and the diffusion tests reveal that the membranes function by a combination of size and charge selection, exhibiting an anion-selective behavior. Tetracycline antibiotic diffusion investigations have been shown to have rejection rates over 95% thus indicating the potential use of the graphene-based membranes for emerging contaminants removal.

Moreover, Pijuan *et al.*[55] recently reported the impact of graphene oxide on the process of anaerobic digestion of waste activated sludge *via* the use of biochemical methane potential tests. The use of graphene oxide resulted in enhanced efficacy in eliminating a wide range of pharmaceutical compounds, including those that exhibit resistance to natural degradation processes, such as diclofenac, azithromycin, and carbamazepine. The findings of the research revealed graphene oxide as a viable substitute for other carbon-based substances used in the anaerobic digestion procedure, with the aim of augmenting the elimination of medicinal compounds. Further research is required to investigate the transformation processes of each compound's transformation products *via* the use of kinetic monitoring.

Besides that, Mantovani *et al.*[56] revealed the considerable potential and adsorption characteristics of amino acid-covalently bonded graphene oxide nanosheets in the removal of emerging pollutants including pharmaceuticals in aqueous systems. The authors revealed that there was an amino acid loading of around 5% and a modest decrease in oxygen levels from 27% to 14–20%. These observations were linked to the adsorption selectivity and the samples under investigation were subjected to comparison with unmodified graphene oxide, reduced graphene oxide, graphene oxide-lysine, and the reference sample graphene oxide-sodium hydroxide. The various forms of modified graphene oxide (GO) have a greater ability to adsorb bisphenol A (BPA), benzophenone-4 (BP4), and carbamazepine (CBZ) compared to both unmodified GO and reduced graphene oxide (rGO). Furthermore, this adsorption process is shown to take place during the first hour of contact time. The maximal adsorption capacity, as determined from the adsorption isotherms, exhibited a strong correlation with the loading of amino acids. The molecular dynamics simulations revealed that the modified graphene oxide (GO) exhibited greater interaction energies compared to the unmodified GO. This may be attributed to the increased van der Waals and hydrophobic contacts

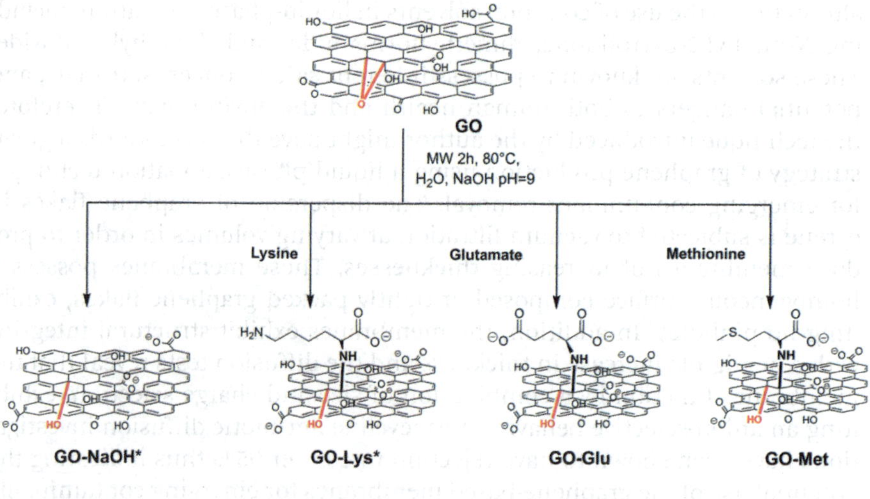

Figure 6.12 Synthetic pathway to produce amino acid-modified graphene oxide. Reproduced from ref. 56 with permission from the Royal Society of Chemistry.

between the contaminants and the amino acid side chains present on the surface of the nanosheet. These GO-amino acids were synthesized *via* microwave-assisted epoxide ring opening reaction, as presented in Figure 6.12.

Graphene and functionalized graphene materials have shown significant efficacy in the removal of emerging pollutants. However, the primary limitations associated with graphene-based nanomaterials are to their elevated cost and challenges pertaining to their reusability. The potential impossibility of reusing the nanomaterial may be attributed to its significant electrostatic interactions thus further investigation of the reusability of graphene-based materials as adsorbents for the removal of emerging pollutants are crucially important. As concerns about water scarcity and pollution continue to mount, harnessing the potential of graphene and graphene oxides for water purification can pave the way for sustainable and effective solutions. By unravelling the intricate interplay between these remarkable materials and emerging water pollutants, researchers and engineers can contribute to the development of innovative technologies that ensure the availability of clean and safe water for present and future generations.

6.2.5 Carbon and Graphene Quantum Dots

Carbon quantum dots (CQDs) are quasi-spherical carbon nanoparticles with exceedingly small sizes (1–10 nm), which have accidently been found to be fluorescent during the purification of SWCNT.[57] Whereas graphene quantum dots (GQDs or G-dots), are small fragments with lateral size less than 10 nm graphene layer with extraordinary luminescence properties[58] they have different formation mechanisms, micro-/nanostructures, and properties, while associations can be built among them by changing the graphene layer and carbonization degree. Both CQDs and GQDs exhibit unique optical and electronic properties, such as tunable photoluminescence and efficient light harvesting, making them suitable for eliminating various pathogens, organic dyes, chemicals, and pesticides from water.[59,60] The synthesis of carbon and GQDs has been extensively studied using various approaches and techniques. These include top-down and bottom-up methods involving cleaving or breaking down of carbonaceous materials *via* chemical oxidative agents, microwave irradiation, ultrasound, hydrothermal conditions, laser irradiation, electrical currents, or plasma.[59,61]

6.2.5.1 *Carbon Quantum Dots for Emerging Pollutant Adsorbent*

From coffee ground waste, Muangmora *et al.*[62] prepared CQDs, combined it with titanium dioxide (TiO_2) and coated fiberglass cloths (FGCs) forming CQDs/TiO_2-FGS photocatalysts. This material successfully removed caffeine with an efficiency of 82% after 9 hours of irradiation. One of the optimized photocatalyst (5CQDs/TiO_2) was then applied to remove caffeine and other pharmaceuticals, including carbamazepine, and ibuprofen with residual concentrations lower than reported toxicity levels. The presence of CQDs in the composite was found to obstruct the recombination of photo-excited electrons and holes, enhancing the separation and utilization of charge carriers for the degradation of pollutants which exhibits higher photocatalytic activity, leading to greater removal efficiency.

Congo red and a few heavy metals can be removed from *in situ* doping lignin-derived CQDs on magnetic hydrotalcite producing N,S-CQDs@Fe_3O_4@HTC composite.[63] The catalyst had a mesoporous structure, facilitating the diffusion and mass transfer of pollutant molecules inside the catalyst which showed high efficiency in the UV degradation of Congo red over a wide pH range (3–11) and even in the presence of high NaCl content (100 g L^{-1}). Furthermore, excellent removal efficiency was observed for removing Cu^{2+} (99.90%) and Cd^{2+} (85.08%) simultaneously. The main active species governing CR degradation were $^{\bullet}OH$ and $^{\bullet}O^{2-}$, as demonstrated by ESR analysis and free radical quenching experiments.

Kumar *et al.*[64] investigated the enhanced photocatalytic removal of Cd(II) from aqueous solution using Bi/S co-doped carbon quantum dots. The doping of carbon dots with Bi and S reduced the band gap, making it suitable for photocatalytic conversion of Cd(II) to Cd(0). The effect of pH, time, and adsorbent dose were found to be significant parameters for Cd(II) removal. Different scavenger solvents were tested, and it was concluded that solvents combining faster with the holes helped reduce the recombination of electron–hole pairs, enhancing Cd(II) removal. Adsorption isotherm models (Langmuir, Freundlich, and Temkin) were fitted to the data, with Langmuir and Temkin isotherms showing good fit and indicating a combination of chemisorption and energy of electrons for Cd(II) removal. Thermodynamic parameters (enthalpy change and entropy change) were estimated, showing that the process was spontaneous and exothermic.

6.2.5.2 Graphene Quantum Dots for Emerging Pollutant Removal

Pristine GQD have limited applications which require modulation to enhance their properties such as through doping, conjugation, and surface modification.[65–67] The addition of GQD in a mixture of nanocomposites and doped-GQD, especially as photocatalysts, significantly improved their capacity to degrade water pollutants. For instance, methyl orange and tetracycline can be degraded with GQDs@CNBP heterojunction, with a photodegradation efficiency of 98% for MO and 97% for TC.[68] The synthesized nanocomposite which consists of graphitic carbon nitride (CN), black phosphorus (BP), and GQD showed higher photocatalytic activity compared to individual components and binary heterojunctions. The degradation of intermediates in wastewater were not detected, indicating successful photocatalytic degradation using a complex-type-II mechanism with the farthest charges on the band edges, generating oxidative species that played a vital role in the reactions, resulting in high kinetics for the GQDs@CNBP heterojunction.

Using hybrids of F-doped TiO_2 and N-doped graphene quantum dots (F-TiO_2@N-GQDs), several pollutants such as methylene blue (90% degradation), ciprofloxacin (62% degradation), and naproxen (60% degradation) demonstrated excellent to good degradation in up to 15 minutes except 4-chlorophenol, which exhibit 59% degradation in 30 minutes.[69] They introduced the F-doping on TiO_2 to decrease the band gap, recombination rate, and induce the formation of charge capture centres resulting in enhanced surface characteristics and electronic properties of the oxide material. In contrast, N-doped GQDs function as repositories for electrons, therefore enhancing the longevity of electron–hole pairs in titanium dioxide (TiO_2) through the reduction of charge recombination. The incorporation of these components enhances the photocatalytic efficiency of the hybrid system.

Rabeie and Mahmoodi[70] combined GQD and TiO_2 with MOFs, forming the ternary TiO_2/GQD/ZIF-8 nanocomposite to degrade doxycycline and dyes (Direct Red 23, Direct Red 80) under LED light. Increased GQD doping enhances photocatalytic activity, but further increases to 0.2 g aggregate the GQD layer on TiO_2 surface, hindering effective light irradiation. The degradation mechanism proposed that the presence of oxygen vacancies in TiO_2 and the sensitization effect of GQD contribute to an increase in the production of oxygen radical species, such as superoxide anion radicals (O_2^-) and hydroxyl radicals (OH), which are essential to the degradation process (see Figure 6.13). Then, light exposure to the photocatalyst initiates a series of redox reactions in which excited electrons are transferred from the valence band (VB) to the conduction band (CB), producing electron–hole pairs. The holes in the VB are responsible for oxidation processes, while the electrons in the CB interact with O_2 to generate O_2^-. OH is produced when photo holes in the VB react with water. These active species interact with contaminant molecules, resulting in their decomposition.

When a moderate band gap composite, copper tungstate ($CuWO_4$) was coupled with GQD, phenol degraded better under visible light by photocatalysis.[69]

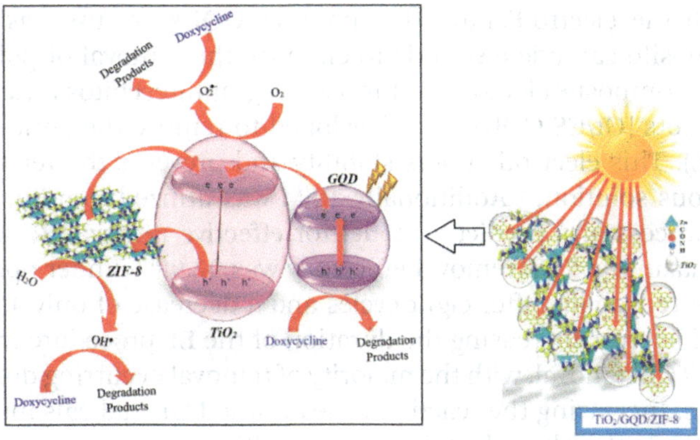

Figure 6.13 Photocatalytic degradation mechanism using GQD.[70] Reproduced from ref. 70 with permission from Elsevier, Copyright 2023.

The 0.5 GCW sample (optimized sample with 0.5 wt% $CuWO_4$ in relation to a fixed quantity of GQDs) exhibited the maximum efficiency of 53.41%, the highest value and the fastest degradation kinetics, with a rate constant of 0.0053 min^{-1}. The enhanced photocatalytic activity might be attributed to the higher charge separation and efficient electron transfer caused by the addition of GQDs to $CuWO_4$ in the composite photocatalyst.

Mahmoud *et al.*[72] prepared a novel supramolecular network consisting of GQD, vitamin B9, and iron(III)-tannic acid complex (GQDs@FA@Fe-TA) as a nanosorbent for the removal of inorganic anionic (Cr(VI)) and organic cationic (malachite green, MG) pollutants from different water samples. The nanosorbent exhibited high removal efficiencies for both pollutants, with 99.5% removal of Cr(VI) at pH 3.0 after 5 minutes and 94.8% removal of MG at pH 6.0 after 30 minutes. The adsorption of both pollutants was exothermic and spontaneous following the pseudo-second order kinetic as well as Langmuir model. After five consecutive adsorption–desorption cycles, the GQDs@FA@Fe-TA supramolecular network demonstrated excellent regeneration stability, with a loss of only 2.1–4.3%. MG removal efficiencies varied between 89.8% and 92.1%, while removal efficiencies for Cr(VI) ranged from 91.2% to 99.4% when tested with other water samples (tap water, sea water, and wastewater).

In a different study, Mahmoud *et al.*[73] successfully developed a 3D nanocomposite structure with multiple functionalities by coupling GQD with metal (Fe–Al) double layered hydroxide (FA-DLH) and intercalated with gemifloxacin antibiotic (Gem) *via* coprecipitation. Under specific pH conditions, this GQDs@Gem@Fe-DLH nanocomposite was able to remove 4-nitroaniline (90.7%) and Pb^{2+} (98.8%). Additionally, it demonstrated antibacterial activity against both Gram-positive bacteria (*E. faecalis*) and Gram-negative bacteria (*E. coli*) with removal percentages of 91.3% and 82.9%, respectively, and achieved 80% removal bacteria from wastewater based on total plate count.

Through the electro-Fenton (EF) process, GQDs are used as part of the nanocomposite cathode electrode to enhance the removal of pollutants. For instance, a composite electrode comprised of graphite, chitosan, and graphene quantum dots (Gr/CS/GQDs) was developed to remove the antibiotic tetracycline (TCL). This electrode offers stability and prevents the release of GQDs into aqueous solution.[74] Additionally, RSM was utilized to optimize the TCL removal process and predict a model for effective removal. Results showed that the maximum TCL removal efficiency was 92.98% under optimal conditions, with reusability after eight cycles and a decrease of only 4.34% in TCL removal efficiency. Increasing the duration of the EF procedure improved the efficacy of TCL removal, with the majority of removal occurring during the first 30 minutes. Increasing the reaction time afforded OH radicals more opportunities to attack, thereby enhancing removal efficiency over time.

Alternatively, α-FeOOH, also known as goethite, was combined with GQDs to enhance the absorption intensity of visible light, promoting the separation of more photogenerated carriers to react with H_2O_2, leading to improved degradation efficiency of contaminants like ciprofloxacin (CIP).[75] The GQDs/α-FeOOH + H_2O_2 + Vis system demonstrated a 93.73% degradation efficiency of CIF, with a higher efficiency using a lower amount of H_2O_2. The combination of α-FeOOH and GQDs generated reactive species, including hydroxyl radicals, superoxide radicals, and singlet oxygen, responsible for CIP degradation. The nanocomposites also demonstrated broad-spectrum UV-vis-IR responsiveness, durability, and recyclability up to three cycles as well as showing acceptable catalytic efficacy in different surface water matrices.

6.2.6 Carbon Nanotubes and Nanofibers

Carbon nanotubes (CNTs) are elongated nanostructures made of carbon atoms in sp^2 hybridization arranged in a cylindrical form or tube which become a great material to reduce water crises and water pollution problems.[76] Generally, they can be categorized as single-wall carbon nanotubes (SWCNTs) and multi-wall carbon nanotubes (MWCNTs) depending on the number of concentric layers present in their structure.[77] Additionally, the arrangement of carbon atoms can either be zigzag, armchair, or chiral structure[78] as illustrated in Figure 6.14. CNTs exhibit unique properties such as excellent mechanical strength, thermal, and electrical conductivities making them highly valuable in wastewater treatment as sorbent, catalyst, or membrane.[79] The most well-known syntheses to produce CNTs are arc-discharged,[80] laser ablation[81] and chemical vapour deposition (CVD) methods.[82]

CNTs are a hydrophobic material very prone to aggregate owing to van der Waals interactions between the tubes, which results in poor solubility, dispersibility, and difficult manipulation in water and polymeric solvents. Therefore, functionalizing the CNT surface reduces the agglomeration and increases the interactions with matrix materials as well as its efficiency. Functionalization or modification involved the introduction of more polar organic groups, biopolymers, proteins, and metals.[84]

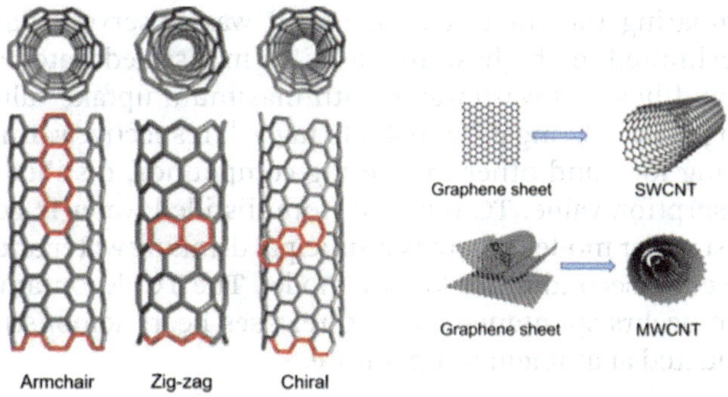

Figure 6.14 Different formation of carbon nanotubes. Reproduced from ref. 83 with permission from Elsevier, Copyright 2023.

6.2.6.1 Application of CNTs in Removing Pharmaceutical Contaminants

Numerous studies have been adopted to remove pharmaceutical contaminants (PC) using CNTs and functionalized-CNTs. Wang *et al.*[85] investigate the effectiveness of CNTs and carboxylated-CNTs (COOH-CNTs) as membrane filtration to combine with electrochemical technology with the aim of enhancing separation performance in removing antibiotics and antibiotics resistance genes (ARG) from wastewater. Of two electrochemical membrane studies, functionalized CNT with carboxylic group (COOH-CNTs) create more defects and active sites, hence, enhance the electrochemical oxidation properties. This led to the increase in antibiotic removal efficiency by 82% and a decrease in ARG concentration by 1.85 log. The degradation pathways of antibiotics were analysed, and it was found that the hydroxylation and ring opening reactions were the main pathways.

The feasibility of optimizing the green absorption material, corncob-based biochar using microorganisms, CNTs, and iron oxide were successfully prepared by Fan *et al.*[86] to produce multilayer porous biochar. Utilization of microorganisms were found to reduce agglomeration problems by CNTs whereas introduction of CNTs were found to further improved the surface area and pore structure. As a result, the maximum adsorption capacity of the modified biochar reached 72.59 mg g^{-1}, and the specific area was greatly increased to 237.51 m^2 g^{-1}. The removal of oxytetracycline (OTC) was influenced by various mechanisms, including chemisorption, multilayer interactions, particle diffusion, ligand exchange, π–π bonding interactions and others. These findings suggested that the modified biochar has potential as a green material for large-scale OTC removal from wastewater.

Foroutan *et al.*[87] explored the removal efficiency of tetracycline (TC), from different aqueous solutions using synthesized adsorbents consisting of CNT alone, and modified CNTs ($CNT/MnFe_2O_4$ and $CNT/\beta\text{-}CD/MnFe_2O_4$).

When comparing the three adsorbents, it was observed that CNT/β-CD/$MnFe_2O_4$ exhibited the highest uptake of TC in distilled water, urban drinking water and hospital wastewater, with maximum uptake values recorded as 89.53 mg g^{-1}, 55.02 mg g^{-1} and 40.36 mg g^{-1}, respectively. Due to the lack of co-existing ions and other interfering compounds, distilled water has a greater absorption value. TC removal from distilled water is governed by a pseudo-first-order model, whereas municipal drinking water and wastewater follow a pseudo-second-order kinetic model. The TC decontamination process, which occurs spontaneously and releases heat, demonstrates efficacy when conducted at ambient temperature.

6.2.6.2 *Application of CNTs in Removing Industrial Contaminants*

Abdullah *et al.*[88] oxidized multiwalled CNTs (MWCNTs) and functionalized them with Fe_3O_4 and poly-NIPAM polymer to create a nanosorbent that is capable of removing kerosene with 95% efficiency. At a pH of 3.5 and 40 °C, the nanosorbent exhibited maximum adsorption capacity of 8.1 g g^{-1}. The P-NIPAM/Fe/MWCNT nanocomposite can adsorb kerosene through hydrogen–hydrogen and van der Waals interactions.

To enhance the catalytic performance of Bisphenol A (BPA) degradation, a dual-cathode electro-Fenton (EF) system was employed.[89] The system consists of a gas diffusion electrode for H_2O_2 production and an active cathode modified with Fe/S doped multi-walled carbon nanotubes (Fe/S-MWCNT) for pH adjustment and iron species supply. This unique combination resulted in a synergistic effect up to 90.3% of synergy factor an a 12.4-fold increase in a catalytic activity compared to single cathode alone. Moreover, it facilitates a broad range of pH applications within 60 minutes from 9.0 to 3.4 without the need for additional reagents, high reusability after five cycles (85.6% BPA removal), and produce low toxic side products, indicating that the constructed dual EF system can substantially reduce the toxicity of BPA wastewater.

Alternatively, BPA and *p*-chlorophenol adsorption and degradation were successfully accomplished using hydroxylated-MWCNTs with copper–nickel ferrite (HCMWNTs/Fe_3O_4) to improve adsorption capacity and material recovery.[90] The results demonstrated that HCMWNTs/Fe_3O_4 has a greater potential for adsorbing BPA (113.0 mg g^{-1}) in comparison to *p*-chlorophenol (41.6 mg g^{-1}). The degradation process was found to occur at a wider pH range (4–10) when KPS was used as oxidant. This resulted in the breakdown of 90% BPA and 88% of the *p*-chlorophenol within 3 and 2 hours, respectively. After three cycles, the adsorbent is still reusable with good percentage clearance.

6.2.6.3 *Application of CNTs in Removing Inorganic Contaminants*

CNTs have been utilized in a variety of ways to combat emergent inorganic pollutants. For instance, incorporating a sustainable approach by using immobilized bioremediation for the treatment of toxic substances in

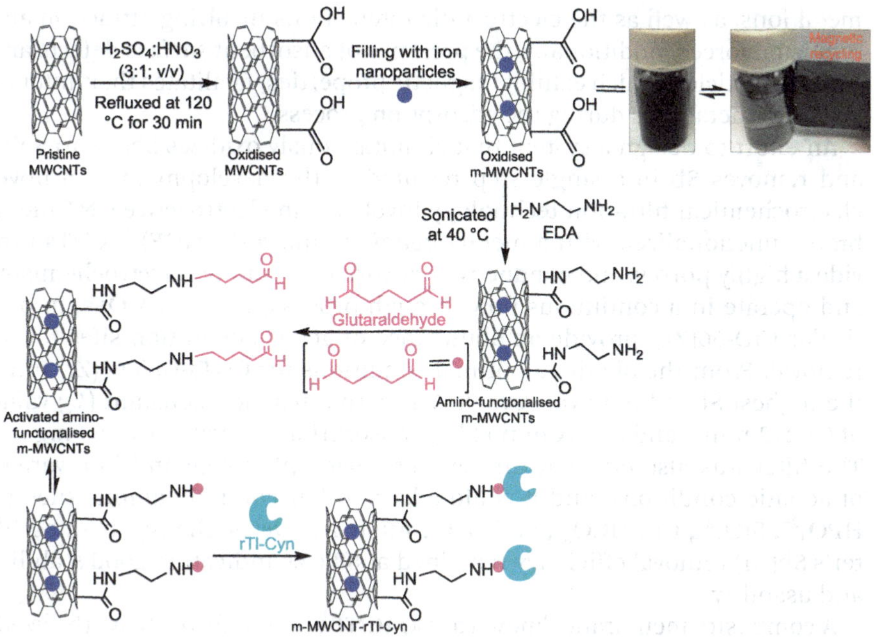

Figure 6.15 Schematic illustration of the synthesis process to produce m-MWCNTs and their activation for rTl-Cyn immobilization.[91] Reproduced from ref. 91 with permission from Elsevier, Copyright 2019.

industries.[91] The cyanate hydratase (rTI-Cyn) which was capable of degrading cyanate was functionalized with iron oxide (m-MWCNTs) prior to covalent bonding to enhance recovery and provide a stable enzyme-adsorbent complex (see Figure 6.15). After a month, the prepared m-MWCNT-rTl-Cyn composite retained almost 100% initial activity after a month whereas the activity of the system remained above 94% after 10 cycles. Intriguingly, the enzyme is also capable of capturing other heavy metals such as chromium, iron, lead, and copper by 39.31%, 35.53%, 34.48%, and 29.63%, respectively, in addition to cyanate (84%). These findings indicate the dual functionality of the immobilized enzyme, thereby presenting opportunities for the development of biocatalytic processes that are more economically viable, efficient, and environmentally sustainable.

Another sustainable strategy was achieved by Abu Bakar *et al.*[92] utilizing waste cooking palm oil (WCPO) as a carbon source to produce CNTs. Subsequently, these CNTs are encapsulated with Fe nanoparticles to create effective adsorbent materials capable of removing Fe(II), Zn(II), and Cu(II) metal ions. Herein, through a modified thermal chemical vapor deposition (TCVD) process, CNT material with 89% purity was attained. Data showed that the synthesized material can remove metal at different pH with good to high removal rate (80–99%). The proposed adsorption mechanism consists of a chemical reaction between the functional groups of the adsorbents and metal ions, as well as the electrostatic interactions involving attraction and repulsion forces. Additionally, the presence of adsorbent surface defects and Fe nanoparticles, which exhibit magnetic properties, facilitates their interaction with metal ions during the adsorption process.

An effort to design a system that simultaneously oxidizes Sb(III) to Sb(V) and removes Sb in a single step resulted in the development of a novel electrochemical filtration technology involving an electroactive CNT membrane functionalized with a metal–organic framework (MOF).[93] CNTs provide a highly porous and conductive network that support electrochemistry and operate in a continuous flow-through process, whereas MOFs, in particular UiO-66(Zr), provide an abundance of active adsorption sites for Sb removal. From the obtained optimal conditions for CNTs-UiO-66(Zr) filter, the highest Sb(III) removal gave 90.5%, with a reaction constant (k_1) value of 0.0272 min^{-1} and a maximum Sb_{total} adsorption capacity of 11.68 mg g^{-1}. The filter was also efficient throughout a wide pH range and in a variety of aquatic conditions, and it could adsorb other common anions such as HPO_4^{2-}, SiO_3^{2-}, Cl^-, HCO_3^-, and NO_3^-. After five cycles, the regenerated filter's Sb(III) removal efficiency remained at 88.6%, indicating good stability and usability.

A composite membrane, known as CQACNT-6, was effectively synthesized, exhibiting a notable capacity for removing heavy metals such as Pb (89.53%), Ni (90.42%), Cu (91.43%), and Zn (91.86%).[94] CQACNT-6 exhibits distinct characteristics when compared to the pristine (CQACNT-0), mostly owing to the inclusion of poly(ether sulfone) (PES) polymer inside the CQACNT-6 membrane, which contributes to the enhancement of surface properties, thus leading to the formation of supplementary channels for water. CQACNT-6 can be synthesized by the reaction between bis[3-(trimethylsilyl)propyl]amine and MWCNT-OH, resulting in the formation of crosslinked aminosilane CNT (CASCNT). Subsequently, the quaternary ammonium group (GTMAC) was introduced using an epoxide ring-opening process, followed by its combination with PES polymer to provide the optimized membrane. The CQACNT-6 was able to retain up to 87% of its initial pure water permeability and 94% of its initial Cu^{2+} rejection after nine cycles.

Marszałek *et al.*[95] investigate the adsorption potential of the thermally modified bentonite and bentonite-CNTs composites, as well as their efficacy in removing zinc, lead, benzotriazole (BZT) and antracene (ANT) from rainwater. Of the five bentonites tested, Ben100 + SWCNT and Ben100 + SWCNT-OH displayed the largest specific surface area and greater microporosity which exhibited a removal percentage of more than 90% for all chemicals tested. The fitting degree showed that the Freundlich model was the most suitable for BZT and ANT adsorption on Ben100 and Ben100 + SWCNT, while Langmuir provided the best fit for Ben100 + SWCNT-OH. During adsorption of these compounds, the time required to reach equilibrium was very short (30 min) for both composites.

Using molecular dynamics (MD) simulation, Ghasemi *et al.*[96] investigate the impact of functionalizing nanotubes with –COO– and COOH-functional groups, as well as the nanotubes' ability to absorb lead ions in two distinct concentrations. To better evaluate the adsorption process, a set of descriptors, including interaction energies, radial distribution function, *etc.*, are calculated. By modifying the surface of CNT and BNNT with functional groups, the absorption efficacy of nanotubes was shown to be significantly enhanced. Due to the negative charge of the –COO–group, which adsorbs positively charged lead ions, the –COO–functional group has substantially enhanced the adsorption efficacy of nanotubes. This study's findings provide significant insight into the mechanisms of lead adsorption on the surface of nanotubes.

6.2.6.4 Application of CNTs in Removing Pesticide Contaminants

Modification with carbon nanotubes (CNTs) have demonstrated notable efficacy as sorbents for phosphate. For instance, the study conducted by Zong *et al.*[97] reported that La-COOH-MWCNTs exhibited a phosphate sorption capacity of 48.02 mg g^{-1}. In contrast, Huang *et al.*[98] observed that chitosan-MWCNTs had a lower phosphate sorption capacity of 36.4 mg g^{-1}. The adsorption capacity for phosphate (198 mg g^{-1}) was seen to improve when MgO nanoflakes were modified with MWCNTs as this can provide an additional active site for phosphate to engage with metal *via* chemisorption, as supported by kinetic and isotherm models.[99] The presence of co-existing ions such as chlorine (Cl^-), nitrate (NO_3^-) ions, and humic acid did not have any detrimental effects, except for sulphate (SO_4^{2-}) and carbonate (CO_3^{2-}), which exhibited reduced efficacy in the removal of phosphate. Another study has shown that electroactive carbon nanotube (CNT) filter functionalized with cerium-based metal organic frameworks (Ce-MOF) can selectively remove phosphate contaminant in a pH range from 3–11. It also exhibited a negligible effect in three types of water samples with the best adsorption capacity found to be 22.41 mg g^{-1}.[100]

Multi-walled carbon nanotubes (MWCNTs) bearing hydroxyl (-OH) or carboxyl (-COOH) surface groups and subsequently doped with magnetite (Fe_3O_4), a metal oxide, have shown the ability to remove 2,4-dichlorophenoxyacetic acid (2,4-D) and atrazine.[101] These organic pesticides are known to be very resistant to degradation in aqueous conditions, making their removal difficult. The MWCNT-OH-Mag composite had the maximum adsorption capacity for 2,4-D and atrazine, with values of 51.4 and 47.7 mg g^{-1}, respectively due to the higher concentration of Fe atoms doped onto the surface of the MWCNT-OH-Mag adsorbent. The presence of these iron (Fe) atoms allows for a direct interaction between the insecticides and the metal, which improves adsorption. The best pH for atrazine removal was six, while the optimal pH for 2,4-D removal was two. These results correspond with the predictions made by molecular modelling, indicating that herbicides demonstrate a lack of electrical charge when exposed to the optimal pH values described earlier.

6.2.6.5 Application of CNF for Emerging Pollutant Removal

CNFs are ultrathin, one-dimensional carbon hollow cylindrical nanostructures which consist of graphene layers that are arranged as stacked cones, cups, or plates with diameters typically in the range 50–500 nm[102] as illustrated in Figure 6.16. When CNFs with graphene are formed into cylinders, they are referred to as CNTs. CNFs are considered perfect substitutes for CNTs due to their low price and availability.[103] Generally, CNFs can be prepared using chemical vapor deposition (CVD) and electrospinning. CVD allows for precise control of the diameter, crystallinity, and orientation of the CNFs by adjusting the synthesis conditions. It can be synthesized by catalytic CVD of hydrocarbons or carbon monoxide over a metal surface or metal alloy catalyst, with the reaction taking place at temperatures in the range 500–1500 °C.[102] Meanwhile, electrospinning produces inexpensive CNFs from polymer solutions or polymer melts (*e.g.*: polyacrylonitrile (PAN), polyimide (PI), poly(vinyl alcohol) (PVA), and polyvinyl chloride (PVC)) as precursors. It involves an electric force to draw charged threads of the polymer, resulting in fibers with diameters in the nanometer range.[104] CNFs have unique properties that make them effective for removing water pollutants. They have high porosity and permeability, allowing for high capacity and throughput sorption of contaminants in water. Recently, their applications have been

Figure 6.16 Cylindrical nanostructures of CNF. Reproduced from ref. 105 with permission from Elsevier, Copyright 2019.

extended to be used for the removal of personal care products, insecticide, pharmaceutical residues, and organic contaminants.

A few paraben derivatives were efficiently extracted from different water samples using a green approach solid-phase extraction (SPE) material consisting of biodegradable PEI (poly(ethyleneimine)) polymer, CNF, and cellulose paper.[106] The sorbent material, known as PEI-CNFsCP, shows a notable capacity to extract parabens when the dimension of the sorbent is 1 × 2 cm and achieves a maximum adsorption equilibrium at 15 min between parabens and sorbent material. Samples tested from river water, lake water, home sewage water, and drinking water yielded favorable results in terms of spiking recovery and precision. The cellulose paper-sorbent material coated with PEI-CNFs employed in this technique exhibits limited reusability and susceptibility to degradation. Additionally, the meticulousness of the batch-to-batch production process is of utmost importance. Nevertheless, this method offers a cost-effective, environmentally friendly approach for the adsorption, extraction, and analysis of parabens in water samples of an environmental nature.

Two water soluble pharmaceutical residues, dextromethorphan (Dex) and guaifenesin (Gua) can be eliminated by oxygenated-CNFs with adsorption capacities of 172.4 mg g^{-1} and 250.0 mg g^{-1}, respectively.[107] It also demonstrated significant potential in removing these pollutants from various water sources, achieving removal efficiencies of 96% for bottled water, 94% for tap water, and 94% for Red Sea water. In a different study, tetracycline hydrochloride (TC), diclofenac (DCF), and metoprolol (MT) was degraded through a photocatalytic process.[108] The sorbent composed of surface hydroxylated carbon nitride (g-C_3N_4), combined with Co_3O_4 through an *in situ* growth method to form 0D/1D Co_3O_4-NTCN nanocomposites, which exhibited a unique structure with a large surface area and short diffusion distances, enhancing the photocatalytic performance compared to Co_3O_4 quantum dots (QDs) and NTCN alone. As a result, the photodegradation ratios of tetracycline hydrochloride (TC), diclofenac DCF, and metoprolol (MT) reached 97.3%, 88.9%, and 63.2% respectively within 60 minutes. The adsorbents effectively trapped and degraded the PPCPs, with O_2^- and OH identified as the major radicals involved in the photocatalytic degradation.

Various electrospun nanofiber mats (ENMs) were fabricated to absorb six polar neonicotinoid insecticides in aquatic environments.[109] Different ENM formulations were tested including polyacrylonitrile (PAN) or carbon nanofibers (CNF; carbonized from PAN), with additives including carbon nanotubes (CNTs; with and without surface carboxyl groups), the cationic surfactant tetrabutyl ammonium bromide (TBAB), and/or phthalic acid (PTA; a CNF porogen). Of these, a combination of NF-CNTs and TBAB in PAN increases logKENM-W values in a manner consistent with the enhancement of PAN nanofiber uptake observed in ENMs containing NF-CNTs and TBAB alone. This could be the consequence of the introduction of two distinct types of surface sites that contribute to neonicotinoid uptake *via* distinct mechanisms.

6.2.7 Carbon Aerogels

In recent years, there has been a significant surge of interest and research directed towards three-dimensional (3D) macrostructures based on graphene. These efforts are aimed at addressing environmental concerns in a sustainable manner. Among the promising 3D graphene-based structures are carbon aerogels. These macrostructures possess a range of distinctive attributes that render them highly attractive for diverse applications, spanning energy storage, catalysis, filtration, and actuators.[110]

The initial production of carbon-based aerogels involved the carbonization of organic aerogels derived from resorcinol formaldehyde.[111] While the carbonization of organic aerogels remains a commonly employed method, numerous researchers have explored alternative approaches for creating carbon aerogels, including freeze drying and the sol–gel technique. These carbon aerogels are crafted by assembling two-dimensional graphene-based nanomaterials, which incorporate a variety of derivative materials such as graphene nanosheets, graphene oxide, and reduced graphene oxide.

Carbon aerogels stand out due to their substantial specific surface area and versatile surface chemistry. Furthermore, their capacity for rapid electron conduction and efficient mass transport adds to their appeal.[112] This combination of features makes them particularly effective in the removal of a wide array of emerging water pollutants, including pesticides, herbicides, and pathogens.

6.2.7.1 *Carbon Aerogels in Removing Pharmaceutical and Personal Care Product Contaminants*

The waste originating from pharmaceuticals and personal care products has gained prominence as a notable water contaminant in recent times. This form of contamination is due to the inappropriate dumping or release of pharmaceutical items, their byproducts, and associated chemicals into the natural surroundings, especially water sources. In response to this challenge, derivatives of carbon aerogels have emerged as a highly promising option for addressing the problem.

Tetracycline has been extensively used in pharmaceutics application as an antibiotic due to its broad spectrum. Due to that, tetracycline is the most common and representative pollutant in pharmaceuticals and personal care products. Many researchers have developed carbon-based absorbents, especially aerogel macromolecules, to treat this emerging pollutant. Shen's research teams have successfully generated three-dimensional macrostructures using a self-assembly approach, combining two-dimensional graphene oxide nanosheets with one-dimensional carbon nanotubes.[113] These assembled structures displayed remarkable adsorption performance for oxytetracycline, with an adsorption capacity of 1729 mg g^{-1}, surpassing that of adsorbents based solely on carbon nanotubes. Additionally, the carbon aerogels were enhanced with various substances to reinforce their capacity to eliminate tetracycline. These modifications included sulfobetaine-chitosan/benzoquinone modified graphene oxide, zeolite-imidazolate graphene oxide, and alginate-graphene oxide. These composite structures exhibited noteworthy tetracycline adsorption capabilities. Besides that, Zhang *et al.* also reported the development of aerogel composed of zeolitic imidazolate framework, chitosan and benzoquinone-modified graphene oxide for the removal of pharmaceuticals, and personal care product contaminants in water. The author revealed that the aerogel possessed excellent properties including reusability, antifouling, bacterial filtration, oil–water separation abilities and exceptional pharmaceuticals and personal care product adsorption performance as presented in Figure 6.17.

Zhou *et al.*[115] and Hiew *et al.*[116] have developed chromium-doped graphene aerogel and graphene oxide aerogel respectively, aiming to investigate their potential for eliminating anti-inflammatory drugs. Optimal conditions led to remarkable sensitivity, with a detection limit in the range 0.006–0.012 ng mL^{-1} for the drugs. The adsorbents exhibited reusability for up to nine cycles without compromising performance, thus offering an economically viable approach and minimal waste generation.

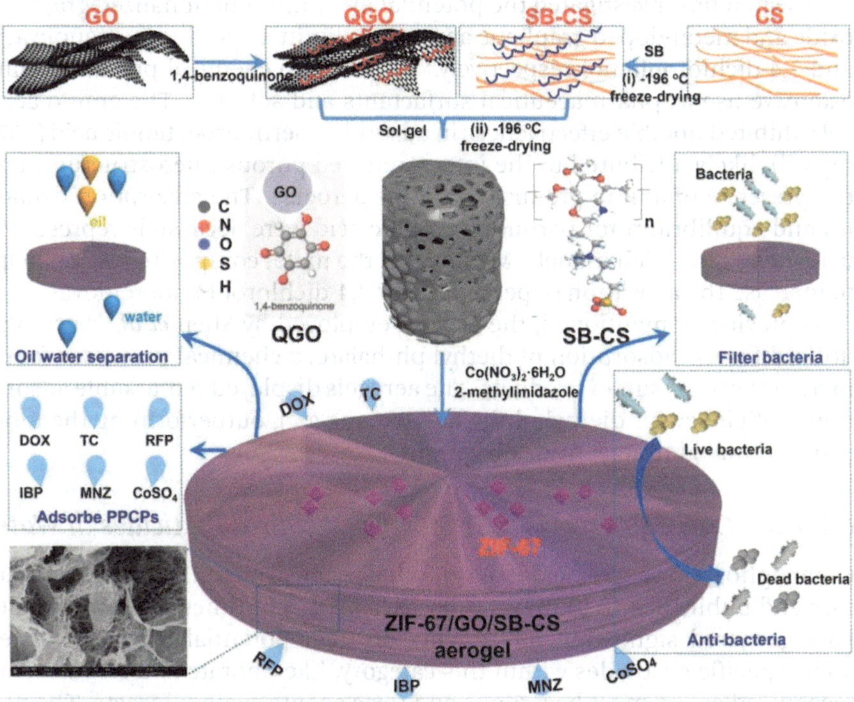

Figure 6.17 Excellent properties of composite aerogel composed of zeolitic imidazolate framework, chitosan, and benzoquinone-modified graphene oxide developed by Zhang *et al.*[114] Reproduced from ref. 114 with permission from Elsevier, Copyright, 2022.

Another emerging pollutant which came from the pharmaceutical field is an antibiotic, sulfamethoxazole. FeOOH graphene polyacrylamide carbonized aerogel was used in neutral electro-Fenton (EF) process to achieve degradation and mineralization of sulfamethoxazole with high removal efficiency.[117] In another study, Wang *et al.*[118] developed a nitrogen-doped graphene aerogel for drug removal from water. The carbon aerogel displayed remarkable catalytic performance in activating peroxymonosulfate for the oxidative degradation of ibuprofen. This resulted in a 44-fold reaction rate improvement compared to graphene aerogel and 8-fold compared to N-doped reduced graphene oxide.

6.2.7.2 *Carbon Aerogels in Removing Chemical Precursors in Water*

Chemical waste precursors, particularly those involved in the production of pharmaceuticals and personal care items, have gained recognition as emerging pollutants, particularly in water systems. Numerous researches have been conducted to explore the potential of carbon aerogels as an adsorption mechanism in addressing this issue.

Research has investigated the potential of amino-functionalized graphene oxide and metal-doped graphene aerogel to eliminate perfluorooctanoic acid and 2,4-dichlorophenol respectively.[119] These two chemical precursors find extensive use as pharmaceutical surfactants and solvents. The amino-aerogel exhibited notable effectiveness in adsorbing perfluorooctanoic acid (1575 mg g^{-1}), likely attributed to the interconnected porous microstructures and the presence of amino groups within the aerogels. The adsorption dynamics and equilibrium for perfluorooctanoic acid were accurately represented using the Freundlich models. Meanwhile, the metal composite aerogel helps to increase the activation of persulfate for 2,4-dichlorophenol removal.

As previously mentioned, the aerogel employed by Shen *et al.*[113] was also utilized for the adsorption of diethyl phthalate, a chemical precursor found in numerous consumer products. The aerogels displayed remarkable adsorption proficiency for diethyl phthalate (680 mg g^{-1}), outperforming the more conventional graphene oxide absorbents.

6.2.7.3 *Carbon Aerogels in Removing Chemical Pesticides in Water*

Organophosphorus pesticides have found extensive application in agriculture and public health to prevent diseases. In recent times, these pesticides have garnered significant interest due to their potential to cause cancer. Some specific pesticides within this category, like phorate and terbufos, are recognized as the most hazardous and toxic agents against insects. The persistent use of these pesticides and their resistance to degradation in soil have resulted in their accumulation in both agricultural and urban soil and water systems. To address this emerging water pollution issue, carbon aerogel stands out as a viable option for the removal of these contaminants.

Liu *et al.*[120] developed carboxylated carbon nanotubes-graphene oxide aerogel through the freeze-drying technique. This aerogel was designed for the purpose of adsorbing glyphosate, an organophosphorus pesticide. By utilizing the Langmuir isotherm model, they determined that the highest adsorption capacity of glyphosate occurred at pH 3, reaching 546 mg g^{-1}. The aerogel demonstrated a notable inclination and capacity for adsorbing glyphosate. The outcomes of the study also indicated that the adsorbent could be reused over 20 cycles while maintaining its effective adsorption capabilities. The equilibrium adsorption procedure conformed to the Langmuir isotherm model, and the intraparticle diffusion model predominantly controlled the adsorption mechanism with exothermic and spontaneous adsorption.

Fan's research group conducted an additional investigation into the elimination of organophosphorus pesticides using carbon aerogel.[121] In this case, they integrated metal iron into the graphene oxide aerogel structure. The team employed a liquid-phase reduction process technique to produce the 3D macromolecules carbon aerogel to activate persulfate, a powerful oxidant, for the breakdown of three distinct organophosphorus pesticides: phorate, terbufos, and parathion, present in water. The composite carbon aerogel demonstrated notably improved catalytic capabilities for activating persulfate to degrade these pesticides when compared to individual iron and carbon aerogel. Impressively, the degradation efficiencies exceeded 99.5% within a mere 5 minute timeframe.

6.3 Conclusion

Carbon-based nanomaterials have emerged as promising candidates for the removal of emerging water pollutants and their unique structural, characteristics, high surface area, and exceptional adsorption properties make them efficient and versatile tools in addressing the challenges posed by a wide range of contaminants in water or aqueous systems. In addition, through various synthesis techniques and functionalization approaches, carbon nanomaterials can be tailored to target specific pollutants, enhancing their selectivity and performance. As our understanding of the interaction between carbon-based nanomaterials and emerging water pollutants continues to evolve, their potential for scalable and sustainable water treatment becomes increasingly evident. Therefore, the use of carbon nanomaterial such as activated carbon nanoparticles, nanodiamonds, fullerenes, graphene and its oxides, carbon and graphene quantum dots, carbon nanotubes and nanofibers, and carbon aerogel should be optimally explored for practical applications. Nevertheless, challenges including long-term stability, scalability, and disposal of spent materials still require further research. Despite these challenges, carbon-based nanomaterials hold great promise for advancing the field of water purification and playing a pivotal role in ensuring safe and sustainable access to clean water in an increasingly polluted world. Continued research and innovation in this area will undoubtedly contribute to the development of efficient and practical solutions for the removal of emerging water pollutants.

6.4 Future Prospects

The future prospects for carbon-based nanomaterials in the removal of emerging water pollutants are exceptionally promising. These innovative materials are poised to play a significant role in addressing the escalating global water pollution crisis. One key avenue of development is the design of nanomaterials with enhanced selectivity and improved adsorption capacity for specific contaminants including pesticides, pharmaceuticals, hormones and steroids, cosmetics, personal care products (PCPs), endoctine disrupting chemicals (EDCs), perfluoronated compounds (PFCs), microplastics and industrial additives. Tailoring the carbon nanomaterials to target a broader range of pollutants will be essential given the ever-evolving landscape of emerging contaminants. Additionally, advancements in scalable and cost-effective production methods are expected, making these nanomaterials more accessible for widespread and practical application. Integration into water treatment processes, such as filtration systems and adsorbent beds, will become increasingly common, contributing to more efficient and eco-friendly water treatment solutions. Moreover, ongoing research into the reusability of carbon-based nanomaterials will reduce the environmental footprint of water treatment processes, enhancing sustainability. Their incorporation into emerging technologies, such as IoT-based water quality monitoring and autonomous water treatment systems, may revolutionize water purification and management. In conclusion, the future of carbon-based nanomaterials for the removal of emerging water pollutants holds great potential. Continued research, development, and collaboration among scientists, engineers, and environmental policymakers are essential to harness the full capabilities of these carbon nanomaterials in safeguarding our precious water resources.

References

1. O. J. Al-sareji, M. Meiczinger, V. Somogyi, R. A. Al-Juboori, R. A. Grmasha, C. Stenger-Kovács, M. Jakab and K. S. Hashim, *J. Environ. Chem. Eng.*, 2023, **11**, 109803.
2. N. Z. Arman, S. Salmiati, A. Aris, M. R. Salim, T. H. Nazifa, M. S. Muhamad and M. Marpongahtun, *Water*, 2021, **13**, 1–31.
3. C. Teodosiu, A. F. Gilca, G. Barjoveanu and S. Fiore, *J. Cleaner Prod.*, 2018, **197**,1210–1221.
4. R. Kumar, M. Qureshi, D. K. Vishwakarma, N. Al-Ansari, A. Kuriqi, A. Elbeltagi and A. Saraswat, *Case Stud. Chem. Environ. Eng.*, 2022, **6**, 100219.
5. H. S. Abbo, K. C. Gupta, N. G. Khaligh and S. J. J. Titinchi, *ChemBioEng Rev.*, 2021, **8**, 463–489.

6. S. Homaeigohar, *Nanomaterials*, 2020, **10**, 295.
7. A. S. El-Kalliny, M. S. Abdel-Wahed, A. A. El-Zahhar, I. A. Hamza and T. A. Gad-Allah, *Discover Nano*, 2023, **18**, 68.
8. P. Kokkinos, D. Mantzavinos and D. Venieri, *Molecules*, 2020, **25**, 1–31.
9. M. Nasrollahzadeh, M. Sajjadi, S. Iravani and R. S. Varma, *Chemosphere*, 2021, 263.
10. S. D. Lakshmi, P. K. Avti and G. Hegde, *Nano-Struct. Nano-Objects*, 2018, **16**, 306–321.
11. R. Baby, M. Hussein, Z. Zainal and A. Abdullah, *Nanomaterials*, 2021, **11**, 3133.
12. A. Gul, N. G. Khaligh and N. M. Julkapli, *J. Mol. Struct.*, 2021, **1235**, 130148.
13. A. Bhatnagar, W. Hogland, M. Marques and M. Sillanpää, *Chem. Eng. J.*, 2013, **219**, 499–511.
14. V. Bernal, L. Giraldo and J. C. Moreno-Piraján, *ACS Omega*, 2020, **5**, 15247–15256.
15. T. J. Al-Musawi, N. Mengelizadeh, M. Taghavi, S. Mohebi and D. Balarak, *Biomass Convers. Biorefin.*, 2023, **13**, 12179–12191.
16. F. Hashemzadeh, M. Ariannezhad and S. H. Derakhshandeh, *Chem. Phys. Lett.*, 2022, **800**, 139656.
17. T. Hata, H. Shintate, S. Kawai, H. Okamura and T. Nishida, *J. Hazard. Mater.*, 2010, **181**, 1175–1178.
18. M. Bilal, M. Adeel, T. Rasheed, Y. Zhao and H. M. N. Iqbal, *Environ. Int.*, 2019, **124**, 336–353.
19. O. J. Al-sareji, M. Meiczinger, R. A. Al Juboori, R. A. Grmasha, M. Andredaki, V. Somogyi, I. A. Idowu, C. S. Kovács, M. Jakab, E. Lengyel and K. S. Hashim, *Sci. Rep.*, 2023, 11933.
20. F. M. de Souza, O. A. A. dos Santos and M. G. A. Vieira, *Environ. Sci. Pollut. Res.*, 2019, **26**, 18329–18342.
21. T. Jayaraman, A. P. Murthy, V. Elakkiya, S. Chandrasekaran, P. Nithyadharseni, Z. Khan, R. A. Senthil, R. Shanker, M. Raghavender, P. Kuppusami, M. Jagannathan and M. Ashokkumar, *J. Ind. Eng. Chem.*, 2018, **64**, 16–59.
22. A. Pandiarajan, R. Kamaraj, S. Vasudevan and S. Vasudevan, *Bioresour. Technol.*, 2018, **261**, 329–341.
23. D. Angın and S. Güneş, *Int. J. Phytorem.*, 2021, **23**, 436–444.
24. N. S. Trivedi, R. A. Kharkar and S. A. Mandavgane, *Arabian J. Chem.*, 2019, **12**, 4541–4549.
25. M. J. Amiri, R. Roohi, M. Arshadi and A. Abbaspourrad, *Environ. Sci. Pollut. Res.*, 2020, **27**, 16983–16997.
26. K. Rambabu, J. AlYammahi, G. Bharath, A. Thanigaivelan, N. Sivarajasekar and F. Banat, *Chemosphere*, 2021, **282**, 131103.
27. D. A. de Freitas, J. A. Barbosa, G. Labuto, R. C. F. Nocelli and E. N. V. M. Carrilho, *Environ. Sci. Pollut. Res.*, 2022, **29**, 79855–79865.
28. A. Kumar, C. Patra, S. Kumar and S. Narayanasamy, *Environ. Res.*, 2022, **206**, 112604.

29. J. Sharma, M. Sharma, S. Nigam and M. Joshi, *Chem. Phys. Impact*, 2023, **6**, 100169.
30. R. Bogdanowicz, *Curr. Opin. Solid State Mater. Sci.*, 2022, **26**, 100991.
31. Y. Zhang, K. Y. Rhee, D. Hui and S. J. Park, *Composites, Part B*, 2018, **143**, 19–27.
32. L. Basso, M. Cazzanelli, M. Orlandi and A. Miotello, *Appl. Sci.*, 2020, **10**, 4094.
33. J. V Macpherson, *Phys. Chem. Chem. Phys.*, 2015, **17**, 2935–2949.
34. R. Bansal, R. Verduzco, M. S. Wong, P. Westerhoff and S. Garcia-Segura, *J. Electroanal. Chem.*, 2022, **907**, 116028.
35. L. Carolina Espinoza, C. Candia-Onfray, J. Vidal and R. Salazar, *Curr. Opin. Solid State Mater. Sci.*, 2021, **25**, 100963.
36. I. M. D. Gonzaga, A. Moratalla, K. I. B. Eguiluz, G. R. Salazar-Banda, P. Cañizares, M. A. Rodrigo and C. Saez, *Sci. Total Environ.*, 2020, **736**, 139536.
37. T. Zhang, Z. Xue, Y. Xie, G. Huang and G. Peng, *RSC Adv.*, 2022, **12**, 26580–26587.
38. P. Brosler, A. V. Girão, R. F. Silva, J. Tedim and F. J. Oliveira, *Front. Mater.*, 2023, **10**, 1020649.
39. T. M. B. F. Oliveira, F. W. P. Ribeiro, S. Morais, P. de Lima-Neto and A. N. Correia, *Curr. Opin. Electrochem.*, 2022, **31**, 100866.
40. H. W. Kroto, J. R. Heath, S. C. O'Brien, R. F. Curl and R. E. Smalley, *Nature*, 1985, **318**, 162–163.
41. A. Astefanei, O. Núñez and M. T. Galceran, *Anal. Chim. Acta*, 2015, **882**, 1–21.
42. J. Singh, P. Nayak, G. Singh, M. Khandai, R. R. Sarangi and M. K. Kar, *C*, 2023, **9**, 3.
43. P. P. Shanbogh and N. G. Sundaram, *Resonance*, 2015, **20**, 123–135.
44. A. Nimibofa, E. A. Newton, A. Y. Cyprain and W. Donbebe, *J. Mater. Sci. Res.*, 2018, **7**, 22.
45. R. Baby, B. Saifullah and M. Z. Hussein, *Nanoscale Res. Lett.*, 2019, **14**, 341.
46. S. Motoc, F. Manea, A. Baciu, S. Vasilie and A. Pop, *Sci. Total Environ.*, 2022, **846**, 157412.
47. N. A. Elessawy, M. Elnouby, M. H. Gouda, H. A. Hamad, N. A. Taha, M. Gouda and M. S. Mohy Eldin, *Chemosphere*, 2020, **239**, 124728.
48. N. Sumi and K. C. Chitra, *Appl. Nanosci.*, 2019, **9**, 2147–2167.
49. A. Kovtun, E. Campodoni, L. Favaretto, M. Zambianchi, A. Salatino, S. Amalfitano, M. L. Navacchia, B. Casentini, V. Palermo, M. Sandri and M. Melucci, *Chemosphere*, 2020, **259**, 127501.
50. Y. Gao, Y. Li, L. Zhang, H. Huang, J. Hu, S. M. Shah and X. Su, *J. Colloid Interface Sci.*, 2012, **368**, 540–546.
51. A. M. E. Khalil, F. A. Memon, T. A. Tabish, D. Salmon, S. Zhang and D. Butler, *Chem. Eng. J.*, 2020, **398**, 125440.
52. M. Kaur, S. Singh, S. K. Mehta and S. K. Kansal, *Molecules*, 2022, **27**, 6956.

53. J. L. Roberts, S. G. Zetterholm, L. Gurtowski, P. U. A. I. Fernando, A. Evans, J. Puhnaty, K. M. Wyss, J. M. Tour, B. Fernando, G. Jenness, A. Thompson and C. Griggs, *J. Hazard. Mater.*, 2023, **458**, 131737.
54. S. N. Sankar, J. Fernandes, M. B. dos Santos, B. Espiña, P. Alpuim, A. G. Díez, S. Lanceros-Mendez, L. Saini, S. Kaushik, G. Kalon and A. Capasso, *Adv. Funct. Mater.*, 2023, **33**, 2214889.
55. O. Casabella-Font, S. Zahedi, M. Gros, J. L. Balcazar, J. Radjenovic and M. Pijuan, *Environ. Pollut.*, 2023, **324**, 121343.
56. S. Mantovani, T. D. Marforio, S. Khaliha, A. Pintus, A. Kovtun, F. Tunioli, L. Favaretto, A. Bianchi, M. L. Navacchia, V. Palermo, M. Calvaresi and M. Melucci, *Environ. Sci.: Water Res. Technol.*, 2023, **9**, 1030–1040.
57. J. Liu, R. Li and B. Yang, *ACS Cent. Sci.*, 2020, **6**, 2179–2195.
58. D. Ghosh, K. Sarkar, P. Devi, K. H. Kim and P. Kumar, *Renewable Sustainable Energy Rev.*, 2021, **135**, 110391.
59. S. Dorontić, S. Jovanović and A. Bonasera, *Materials*, 2021, **14**, 6153.
60. U. A. Rani, L. Y. Ng, C. Y. Ng and E. Mahmoudi, *Adv. Colloid Interface Sci.*, 2020, **278**, 102124.
61. S. Tajik, Z. Dourandish, K. Zhang, H. Beitollahi, Q. Van Le, H. W. Jang and M. Shokouhimehr, *RSC Adv.*, 2020, **10**, 15406–15429.
62. R. Muangmora, P. Kemacheevakul and S. Chuangchote, *Heliyon*, 2023, **9**, e17693.
63. Y. Lu, H. Xu, S. Wei, F. Jiang, J. Zhang, Y. Ge and Z. Li, *Int. J. Biol. Macromol.*, 2023, **239**, 124303.
64. D. Kumar, J. Pandey and A. Kumar, *Mater. Sci. Energy Technol.*, 2024, **7**, 9–18.
65. Z. Luo, G. Qi, K. Chen, M. Zou, L. Yuwen, X. Zhang, W. Huang and L. Wang, *Adv. Funct. Mater.*, 2016, **26**, 2739–2744.
66. C. Tshangana, M. Chabalala, A. Muleja, E. Nxumalo and B. Mamba, *J. Environ. Chem. Eng.*, 2020, **8**, 103930.
67. D. Wang, J. F. Chen and L. Dai, *Part. Part. Syst. Charact.*, 2015, **32**, 515–523.
68. Z. Eroglu and O. Metin, *ACS Appl. Nano Mater.*, 2023, **6**, 7960–7974.
69. I. J. Gómez, M. Díaz-Sánchez, N. Pizúrová, L. Zajíčková, S. Prashar and S. Gómez-Ruiz, *J. Photochem. Photobiol., A*, 2023, **443**, 114875.
70. B. Rabeie and N. M. Mahmoodi, *J. Water Process Eng.*, 2023, **54**, 103976.
71. A. Sarwar, A. Razzaq, M. Zafar, I. Idrees, F. Rehman and W. Y. Kim, *Results Phys.*, 2023, **45**, 106253.
72. M. E. Mahmoud, N. A. Fekry and A. M. Abdelfattah, *J. Mol. Liq.*, 2021, **341**, 117312.
73. M. E. Mahmoud, N. A. Fekry and S. M. S. Mohamed, *J. Water Process Eng.*, 2022, **46**, 102562.
74. M. Ebratkhahan, M. Zarei and N. Arsalani, *Sep. Purif. Technol.*, 2022, **299**, 121663.
75. M. N. Pervez, S. Ma, S. Huang, V. Naddeo and Y. Zhao, *Water*, 2022, **14**, 2260.
76. R. Das, M. E. Ali, S. B. A. Hamid, S. Ramakrishna and Z. Z. Chowdhury, *Desalination*, 2014, **336**, 97–109.

77. S. Iijima and T. Ichihashi, *Nature*, 1993, **363**, 603–605.
78. G. Mittal, V. Dhand, K. Y. Rhee, S. J. Park and W. R. Lee, *J. Ind. Eng. Chem.*, 2015, **21**, 11–25.
79. K. M. Lee, C. P. P. Wong, T. L. Tan and C. W. Lai, *Mater. Sci. Eng., B*, 2018, **236–237**, 61–69.
80. S. Iijima, *Nature*, 1991, **354**, 56–58.
81. T. Guo, P. Nikolaev, A. Thess, D. T. Colbert and R. E. Smalley, *Chem. Phys. Lett.*, 1995, **243**, 49–54.
82. M. Endo, K. Takeuchi, S. Igarashi, K. Kobori, M. Shiraishi and H. W. Kroto, *J. Phys. Chem. Solids*, 1993, **54**, 1841–1848.
83. S. Mishra and B. Sundaram, *Environ. Nanotechnol., Monit. Manage.*, 2023, **19**, 100764.
84. R. Dubey, D. Dutta, A. Sarkar and P. Chattopadhyay, *Nanoscale Adv.*, 2021, **3**, 5722–5744.
85. J. Wang, H. Liu, X. Chen, Y. Li, X. Sha, H. Song, B. Li, Z. Yan and M. Chang, *Front. Chem.*, 2022, **10**, 1–11.
86. Y. Fan, J. Su, L. Xu, S. Liu, C. Hou, Y. Liu and S. Cao, *Environ. Res.*, 2023, **231**, 116307.
87. R. Foroutan, S. J. Peighambardoust, P. Latifi, A. Ahmadi, M. Alizadeh and B. Ramavandi, *J. Environ. Chem. Eng.*, 2021, **9**, 106344.
88. T. A. Abdullah, T. Juzsakova, P. C. Le, K. Kułacz, A. D. Salman, R. T. Rasheed, M. A. Mallah, B. Varga, H. Mansoor, E. Mako, B. Zsirka, A. K. Nadda, X. C. Nguyen and D. D. Nguyen, *Environ. Pollut.*, 2022, **306**, 119372.
89. Z. Chen, Y. Zhang, T. Cao, R. Zhang and K. Yao, *Environ. Res.*, 2023, **231**, 116099.
90. H. Gao, X. Han, R. Wang, K. Zhu and R. Han, *Environ. Res.*, 2023, **231**, 116314.
91. B. Ranjan, S. Pillai, K. Permaul and S. Singh, *J. Hazard. Mater.*, 2019, **363**, 73–80.
92. S. Abu Bakar, N. Jusoh, A. Mohamed, M. Muqoyyanah, M. H. D. Othman, M. H. Mamat, M. K. Ahmad, M. A. Mohamed, M. N. Azlan, N. Hashim, M. D. Birowosuto and T. Soga, *Environ. Sci. Pollut. Res.*, 2021, **28**, 65171–65187.
93. F. Tian, Y. Ren, W. Wu and Y. Liu, *Chemosphere*, 2023, **335**, 139047.
94. R. P. Pandey, M. Ouda, P. Abdul Rasheed, F. Banat and S. W. Hasan, *npj Clean Water*, 2022, **5**, 44.
95. A. Marszałek, G. Kamińska and N. F. Abdel Salam, *J. Water Process Eng.*, 2022, **46**, 102550.
96. Z. Ghasemi, F. Farzad, A. Zaboli and A. Zeraatkar Moghaddam, *Sci. Rep.*, 2023, **13**, 11377.
97. E. Zong, X. Liu, J. Wang, S. Yang, J. Jiang and S. Fu, *J. Mater. Sci.*, 2017, **52**, 7294–7310.
98. Y. Huang, X. Lee, M. Grattieri, F. C. Macazo, R. Cai and S. D. Minteer, *J. Mater. Sci.*, 2018, **53**, 12641–12649.
99. S. Jiang, J. Wang, S. Qiao and J. Zhou, *Sci. Total Environ.*, 2021, **796**, 148907.

100. W. Wu, Z. Zhao, M. Li, W. Zheng, S. You, Q. Wei and Y. Liu, *Chemosphere*, 2022, **303**, 135226.
101. H. A. Pereira, K. da Boit Martinello, Y. Vieira, J. C. Diel, M. S. Netto, G. D. Reske, E. Lorenzett, L. F. O. Silva, T. A. L. Burgo and G. L. Dotto, *Chemosphere*, 2023, **325**, 138384.
102. Y. M. Manawi, Ihsanullah, A. Samara, T. Al-Ansari and M. A. Atieh, *Materials*, 2018, **11**, 822.
103. L. Feng, N. Xie and J. Zhong, *Materials*, 2014, **7**, 3919–3945.
104. T. Wang, Z. Chen, W. Gong, F. Xu, X. Song, X. He and M. Fan, *ACS Omega*, 2023, **8**, 22316–22330.
105. A. Mohamed, in *Carbon-Based Nanofillers and Their Rubber Nanocomposites: Carbon Nano-Objects*, Elsevier, 2018, pp. 243–257.
106. R. R. Pasupuleti, P. C. Tsai and V. K. Ponnusamy, *Chemosphere*, 2021, **267**, 129274.
107. M. R. Elamin, A. O. Elzupir and B. Y. Abdulkhair, *Environ. Nanotechnol., Monit. Manage.*, 2021, **15**, 100397.
108. S. Dai, L. Xiao, Q. Li, G. Hao, Y. Hu and W. Jiang, *Sep. Purif. Technol.*, 2022, **297**, 121481.
109. M. Nagorzanski, J. Qian, A. Martinez and D. M. Cwiertny, *J. Hazard. Mater. Adv.*, 2023, **9**, 100219.
110. S. Chowdhury and R. Balasubramanian, *Prog. Mater. Sci.*, 2017, **90**, 224–275.
111. H. Pröbstle, R. Saliger and J. Fricke, in *Studies in Surface Science and Catalysis*, 2000, vol. 128, pp. 371–379.
112. G. Gan, X. Li, S. Fan, L. Wang, M. Qin, Z. Yin and G. Chen, *Eur. J. Inorg. Chem.*, 2019, **2019**, 3126–3141.
113. Y. Shen, X. Zhu, L. Zhu and B. Chen, *Chem. Eng. J.*, 2017, **314**, 336–346.
114. W. Zhang, T. Huang, Y. Ren, S. Yang, X. Zhao, M. Yuan, J. Wang and Q. Tu, *Carbohydr. Polym.*, 2022, **298**, 120102.
115. Y. Zhou, J. Xu, N. Lu, X. Wu, Y. Zhang and X. Hou, *Talanta*, 2021, **225**, 121846.
116. B. Y. Z. Hiew, L. Y. Lee, K. C. Lai, S. Gan, S. Thangalazhy-Gopakumar, G. T. Pan and T. C. K. Yang, *Environ. Res.*, 2019, **168**, 241–253.
117. Y. Wang, H. Zhang, B. Li, M. Yu, R. Zhao, X. Xu and L. Cai, *Chem. Eng. J.*, 2019, **359**, 914–923.
118. J. Wang, X. Duan, Q. Dong, F. Meng, X. Tan, S. Liu and S. Wang, *Carbon*, 2019, **144**, 781–790.
119. D. Tian, D. Geng, W. Tyler Mehler, G. Goss, T. Wang, S. Yang, Y. Niu, Y. Zheng and Y. Zhang, *Sci. Total Environ.*, 2021, **783**, 147041.
120. K. Liu, M. Li, Z. Zhu, B. Gao, H. Zeng, Y. Ma and L. Fang, *Environ. Sci.: Water Res. Technol.*, 2021, **7**, 714–725.
121. M. Fan, P. Zhang, C. Wang, J. Tang and H. Sun, *J. Hazard. Mater.*, 2022, **428**, 128254.

CHAPTER 7

Synthesis and Functionalization of Metal Oxides for the Removal of Organic Pollutants

KHALID UMAR*, SAIMA KHAN AFRIDI AND ANWAR IQBAL

School of Chemical Sciences, Universiti Sains Malaysia, 11800 Penang, Malaysia
E-mail: khalidumar@usm.my

7.1 Introduction

The survival of all living things depends on water. Even though water covers roughly 71% of the earth's surface, fresh water only contributes about 2.5% of that total volume.[1,2] On the other hand, excessive water consumption in various industries over the past few decades has led to a decrease in the amount of freshwater needed for the existence of both humans and wildlife. Furthermore, the lack of access to safe water resources for billions of people leaves them susceptible to illnesses spread through water.[3] Prior to 2015, it is estimated that nearly 70% of the wastewater produced worldwide was not properly treated, resulting in water pollution through its discharge to nearby natural water bodies. Thus, it is anticipated that 50% of the population will experience clean water shortages by 2025.[4]

By considering these points, there is an urgent need to develop effective, dependable, and economically viable methods to deal with the emerging

Chemistry in the Environment Series No. 14
Advanced Materials for Emerging Water Pollutant Removal
Edited by Pei Sean Goh, Devagi Kanakaraju, Anwar Iqbal and Ahmad Fauzi Ismail

Published by the Royal Society of Chemistry, www.rsc.org

contaminants and address the safety problems caused by them. Many wastewater treatment technologies have been developed during the last few decades. The most well-known methods are coagulation,[5] sedimentation–flocculation,[6] ion exchange, molecular sieves,[7] reverse osmosis,[8] membrane filtration,[9] adsorption processes,[10] ozonation,[11] chlorination,[12] chemical precipitation,[13] and chemical and electrochemical techniques.[14] According to reports, traditional treatment technologies are ineffective for fully degrading recalcitrant pollutants in wastewater and are neither cost-effective nor energy efficient,[15] they can only change the phase of organic molecules and bacteria, not completely remove them. Thus, cutting-edge, advanced treatment techniques are required to eliminate the organic component of wastewater. Over the past two decades, there has been an unprecedentedly large amount of research on wastewater treatment through advanced oxidation processes (AOPs).[16]

In situ synthesis of strong oxidizing agents, primarily hydroxyl radicals, superoxide anion radicals, and photogenerated electron–hole pairs, by AOPs results in the total degradation and mineralization of pollutants in contaminated water.[17] Due to its relatively low cost, high stability, non-toxicity, lack of mass transfer resistance, lack of secondary pollution, operation under ambient conditions, and, most importantly, the potential to decompose recalcitrant organic pollutants into less harmful compounds in a short reaction time, heterogeneous photocatalysis has been recognized as one of the most alluring and potentially effective AOPs.[18–20] In this context, the use of semiconductor-based photocatalysis has drawn considerable attention from a wide range of academic fields since it provides an effective and environmentally friendly solution to environmental issues.

Photocatalysis appears to be an interesting approach for water purification, with the possibility of using sunlight as a sustainable and renewable source of energy. This technology is based on the use of a semiconductor that can be excited by light at an energy higher than its band gap, leading to the generation of electron–hole pairs with high energy that can be involved in redox reactions. In order to improve their electrical and optical properties and increase their photoresponse to visible light, recent research has investigated the chemical makeup of nanoscale semiconductors. Nanomaterials are well suited for use in water purification because of their high activity, high functionalization, large specific surface area, and size-dependent characteristics.[21]

Due to its special qualities of being economical, having good optical activity, great chemical stability, and being non-toxic, metal oxide-based semiconductors are widely used as a photocatalyst.[22] The metal oxide-based nanoparticles have been used in a variety of fields, including medical research, drug delivery, antimicrobial materials, energy storage, and self-cleaning. Moreover, these metal oxides are extensively used for the photodegradation of several chemical contaminants found in wastewater. It encourages the breakdown of organic molecules when exposed to light.

However, pure metal oxides (photocatalysts) have a broad bandgap, which makes them more prone to excitation when exposed to UV light, and a high rate of electron–hole (e−/h+) pair recombination, which results in a relatively low photocatalytic efficiency.

As a consequence, one way to make a photocatalyst absorb visible light is to alter its inner band structure by substituting oxygen in its lattice to reduce the optical energy bandgap. In this regard, a number of methods (structural and chemical changes) have been used to improve the photocatalytic activity of photocatalysts and optimize it, particularly its bandgap energy and recombination rate. Doping these catalysts with metals and/or non-metals is thought to be an efficient way to boost its photocatalytic efficiency because it prevents electron–hole recombination and increases its ability to harvest visible light.[23,24]

7.2 Various Methods to Synthesize Metal Oxides for the Removal of Pollutants

The synthesis of metal oxides for the removal of pollutants has been carried out using a variety of chemical and physical techniques. They can be divided into bottom-up (chemically) and top-down (physically). Decomposing bulk material into tiny pieces and then converting those smaller pieces into the desired nanoscale particles is known as the top-down or destructive technique. It involves a variety of decomposition techniques, including physical vapor deposition, melt mixing, laser ablation, and high energy ball milling (HEBM) *etc.* The atoms or small molecule precursors are built up to become nanoparticles using a bottom-up or constructive approach after the nucleation process. There are a variety of bottom-up techniques, including sol–gel, hydrothermal, microemulsion, electrochemical, thermal decomposition, co-precipitation, and biological ones.[25] There are some benefits, influencing factors, merits, and demerits of these processes, which are compiled in Table 7.1, since the metal oxide property change is largely dependent on these synthetic steps.[26]

7.2.1 Bottum Up Approach

7.2.1.1 Sol–Gel Method

The sol–gel process, in general, entails the conversion of a system from a liquid "sol", which is mostly in colloidal form, into a solid "gel" phase. Using this method, a wide range of ceramic or glass materials can be created, including thin film coatings, ceramic fibers, microporous inorganic membranes, monolithic ceramics and glasses, and extremely porous aerogel materials.[27] In order to prepare the "sol," metal organic compounds like metal alkoxide and inorganic metal salts are typically employed as

Table 7.1 Various methods are employed for the synthesis of metal oxide nanoparticles. Adapted from ref. 26 with permission from MDPI, Copyright 2022.

Method	Highlight	Influencing factor	Merit	Demerit
Sol–gel	This is comprised of hydroxylation and precursor condensation followed by solvent removal or gelling by reaction.	Precursor concentration, temperature, gel nature, pH, stirring rate, reaction kinetics.	Wide reaction temperature use including room temperature (25–200 °C).	Medium yield. Better shape tuning. The reaction may span hours for good quality colloid production.
Hydrothermal	The disintegration of bulk material into nano/finer particles *via* the high-temperature application.	The atmosphere used, temperature and heating rate.	Highly scalable. Good shape tuning. The capability of producing contaminant-free particles.	May be complex to use. High-temperature use. May induce structural disintegration.
Microemulsion	Water immersed in an oil medium, then stabilized molecular surface medium followed by elimination of surfactant and particle purification.	Salt type, pH, type of surfactant consumed.	Fairly low-temperature use (20–80 °C). Time-consuming method (roughly to hours). Better shape control.	Processability may be complex—low yield of colloids.
Electrochemical decomposition	The occurrence of deposition is at the electrolyte interface housing the metal to be deposited along with the conductive metal substrate.	Applied potential, electrode nature, residence electrolyte chemical nature, metal substrate.	Eco-friendly (less chemical consumption). Readily performed under room temperature. Fair shape control.	Expensive instrumentation. May suffer yield control—limited electroactive electrode use.
Thermal decomposition	The disintegration of bulk material into nano/finer particles *via* the high-temperature application.	The atmosphere used, temperature, heating rate.	Highly scalable. Good shape tuning. The capability of producing contaminant-free particles.	May be complex to use. High-temperature use. May induce structural disintegration.

(continued)

Table 7.1 (*Continued*)

Method	Highlight	Influencing factor	Merit	Demerit
Co-precipitation	The reaction of precursor salt in an aqueous solution to a base with a mild oxidant.	Highly dependent on reagent ratio use, stirring rate, pH, and temperature.	Highly scalable. Relatively narrow size distribution. Fewer reaction periods. Simple and can be performed under ambient conditions.	Possible agglomeration. May result in impurity formation. High pH use during the synthesis and purification. Not suitable for precise stoichiometric phase fabrication.
Biosynthesis	Biomaterials are used as reducing agents to yield particles from the appropriate precursors.	Precursor-biomaterial, mole ratio, temperature, incubation time, pH, agitation rate.	Achievable under room temperature. Reduced toxic chemical use.	May be complex to support reproducibility. Low production yield. Time-consuming. Inadequate nanoparticle shape control.
Vapor method	The vapor composition is made unstable relative to the generation of resultant particles.	Temperature, atmosphere, precursor, synthesis time, evaporation–condensation rate.	Reasonable synthesis time (minutes–hours). Highly scalable. Fair shape control. Good material purity.	Not simple to perform. High-temperature consumption.

starting materials. A sol or colloidal suspension is formed by a succession of hydrolysis and polymerization processes. The sol can be processed further to produce ceramic materials in a variety of forms. A wet "gel" will develop once the sol is poured into the mold. The gel can be further dried and heated to create thick ceramic products. Wet gels can be made into aerogel, a very porous and extremely low-density substance, by removing the liquid under extremely critical conditions.[28] Ceramic fibers can be extracted from a sol by adjusting its viscosity to a suitable range. Precipitation, spray hydrolysis, or emulsion methods are used to create ultra-fine and homogenous ceramic powders. Figure 7.1 depicts the finer points of the sol–gel procedure.[29]

There are two different types of sol–gel processes that can be used to create TiO_2 semiconductors, depending on the type of titanium metal precursor: (i) an alcohol-based process where metal alkoxide serves as the initial precursor, and (ii) an aqueous-based process where inorganic metal salt serves as the initial precursor.[30] For alcohol-based processes, TiO_2 can be produced from the metal alkoxide precursors $Ti(OC_2H_5)$, $Ti(OC_3H_7)_4$, and $Ti(OC_4H_9)_4$. Due to the significant electronegativity difference between Ti and O, the metal–oxygen bond seen in these alkoxides becomes highly polar and incredibly reactive. As a result, when water is added, simultaneous hydrolysis and condensation reactions continue, eventually resulting in the development of a gel. $TiOSO_4$ and $TiCl_4$ are crucial precursors for the aqueous-based method. The hydrolysis and condensation process can also be used to prepare aqueous-based sol–gels. The aqueous-based sol–gel technique includes the steps of peptization and precipitation. With the addition of a base, the inorganic metal salt hydrolyzes quickly to form a gelatinous precipitate. The precipitate is then washed to remove extra electrolyte. Re-dispersion of a colloid to inhibit coagulation is known as peptization. The term "peptization" refers to the direct disintegration of a substance into colloidal-sized particles by an additional agent.

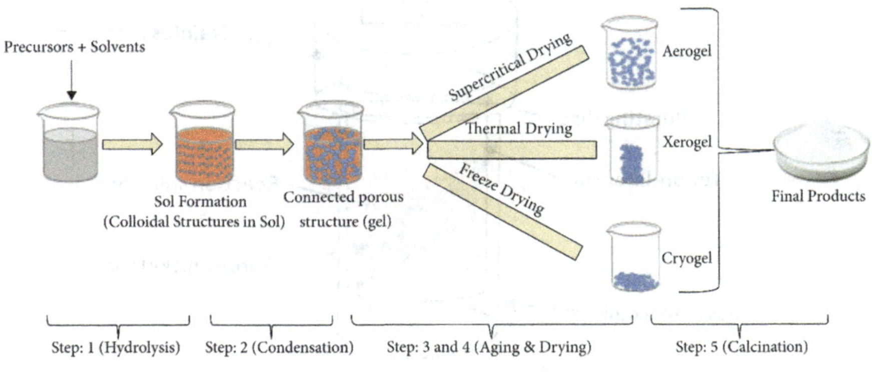

Figure 7.1 Sol–gel method. Reproduced from ref. 29, https://doi.org/10.1155/2021/5102014, under the terms of the CC BY 4.0 license, https://creativecommons.org/licenses/by/4.0/.

A series of chemical reactions occur as a result of the experimental work that has been done by Hasnidawani *et al.*[31] A ZnO colloid should be produced when zinc acetate is completely hydrolyzed with the help of NaOH in an ethanolic solution. The equilibrium between the condensation reaction and hydrolysis led to the generation of the final product. Zinc acetate in the solution is hydrolyzed by the heating, resulting in zinc ions and acetate ions. The hydroxyl groups (–OH) of alcohol molecules form a bond with the zinc ions due to the abundance of electrons in oxygen atoms. A byproduct of the hydrolysis reaction that forms in the presence of water and OH ions is zinc hydroxide acetate. At higher temperatures and over an extended period of time, it can be easily converted into ZnO. Because sodium acetate is water soluble, it might be taken out of the finished product. As a result, the sol–gel process may be used to successfully produce high purity ZnO nano-powder.

7.2.1.2 Hydrothermal Method

Another common method of preparation, particularly for nanostructures that are in powder form, is the hydrothermal method. In order to synthesis a crystalline nanostructure with the right ratio, a solvent was mixed with precursors and necessary agents during the hydrothermal process. A sealed autoclave made of stainless steel that was lined with Teflon and heated in an oven at a particular temperature, time, and autogenous pressure was used during the procedure to seal the mixture. The autoclave's internal pressure and total volume of solution added, depend on its temperature.[32] Figure 7.2 shows the sealed autoclave used in hydrothermal process.

Amorphous TiO_2 powders which were synthesized using the sol–gel precipitation process were used to obtain TiO_2 nanostructures using the hydrothermal method. The hydrothermal system was sealed off from the outside

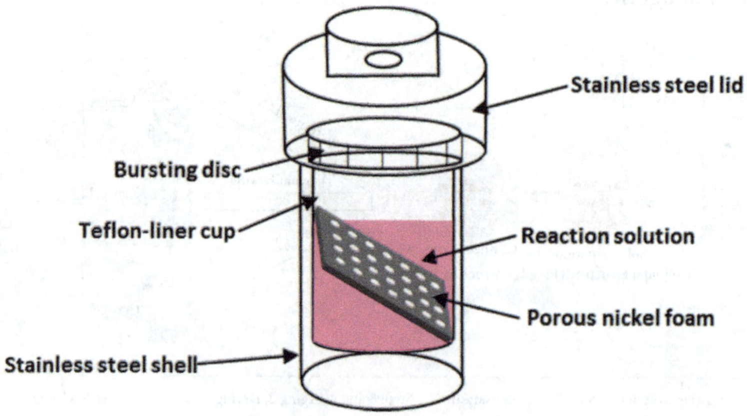

Figure 7.2 Sealed autoclave in hydrothermal process. Reproduced from ref. 32, https://doi.org/10.3390/ma11091560, under the terms of the CC BY 4.0 license, https://creativecommons.org/licenses/by/4.0/.

environment, and hydrothermal processes were permitted to proceed for 36 hours at 130 °C, autogenous pressure, and 250 rpm stirring rate. When an amorphous TiO_2 precursor was used, scanning electron microscopy (SEM) analyses showed that TiO_2 nanofibers rather than nanotubes developed. The powders were repeatedly acid treated by HCl after hydrothermal synthesis. The produced powders were found to be Na-titanate using X-ray diffraction (XRD) examination, and they continued to be Na-titanate even after being subjected to many acidic treatments. By watching methylene blue (MB) solution degrade under UV light, the photocatalytic performance of the powders was assessed.[33]

The hydrothermal process is recognized as an effective synthetic strategy due to its several benefits, including a suitable crystallization temperature, environmental friendliness, controllability of reaction conditions, low energy consumption, and low cost. The base solution employed in the hydrothermal procedure has been observed to have an impact on the one-dimensional titanate nanostructures' final form.

7.2.1.3 Micro-emulsion Method

The micro-emulsion process is one of the most modern and effective ways to synthesize inorganic nanoparticles. Due to their immiscibility, oil and water separate into two phases when combined, each of which contains traces of the other substance. The two phases must be combined, and this requires energy input that would replace the water–water/oil–oil interactions with water–oil association. Bulk oil and water can have an interfacial tension up to 30 to 50 dynes cm^{-1}. Surfactants, which are molecules with surface activity, can be used to get around this. Both hydrophilic and lipophilic moieties can be found in surfactants. They have a propensity to adsorb at the water–oil interface because of this property. The interfacial tension is reduced as a result of the presence of sufficient surfactant molecules, which align and form an interface between the water and the oil. When a small amount of sufficient surfactant is mechanically stirred with the oil and water, a two-phase dispersion is created, with the one phase existing as droplets coated with surfactant that are spread throughout the other continuous phase and this is known as an emulsion. Due to the fact that the droplet sizes of these emulsions range from 0.1 to 1 micron, they have a milky or turbid look. In the case of a hydrophilic surfactant, oil will be emulsified in droplets over a continuous water phase. More lipophilic surfactants behave in the opposite way. In this situation, water will be emulsified in the form of droplets dispersed throughout a continuous oil phase. Although emulsions are kinetically stable, they are eventually thermodynamically unstable and will start to break up again into their two phases. In contrast to the dispersed phases, which will sediment, the droplets will combine. They now disintegrate back into bulk phases of pure water and oil, with some surfactant dissolving preferentially in one of the two.[34] Figure 7.3 depicts an illustration of a reaction using the microemulsion technique.[35]

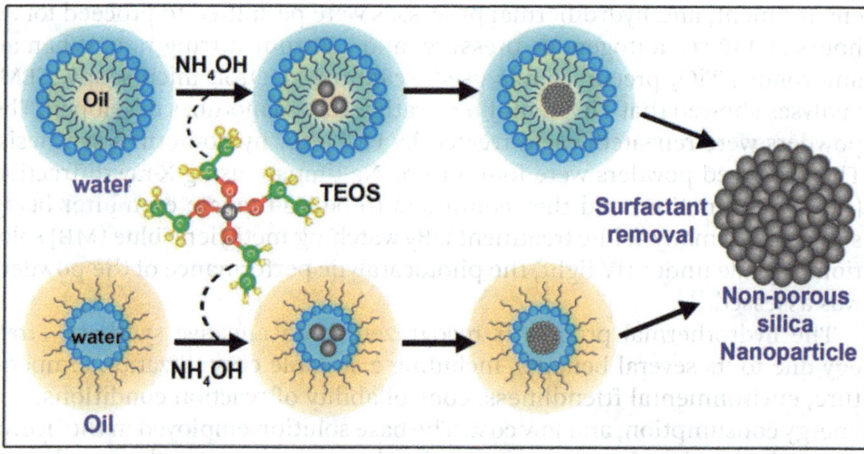

Figure 7.3 Schematic diagram representing microemulsion method for the synthesis of silica nanoparticles (SNPs). Reproduced from ref. 35, https://doi.org/10.3390/pr10122595, under the terms of the CC BY 4.0 license, https://creativecommons.org/licenses/by/4.0/.

The high photocatalytic activity TiO_2 nanoparticles were produced by the hydrothermal surfactant-free microemulsion (SFME) approach at lower temperatures employing oil-in-water SFME systems as templates. According to Cui *et al.*[36] electrical conductivity and UV-visible spectroscopy were used to determine the ethyl acetate (EA)/propan-2-ol/water SFME phase diagram. Transmission electron microscopy, X-ray diffraction, and N_2 adsorption–desorption techniques were used to analyze the produced TiO_2 nanoparticles. The type IV adsorption–desorption isotherms and type H_2 hysteresis loops indicated that they were all classified as anatase phases. According to Brunauer–Emmett–Teller (BET) theory, the TiO_2 nanoparticle's specific surface area increases when the EA content of the SFME lowers. The photocatalytic activity of the TiO_2 nanoparticles is subsequently increased as a result. While the SBET values fall and the crystal size grows as the temperature rises, the photocatalytic activity of the TiO_2 nanoparticles dramatically increases. When prepared at 190 °C, TiO_2 degrades at a rate of 97%, which is much higher than Degussa P-25's rate of deterioration (66%), under the same conditions. This could be because the surfactant-blocking effects of surfactant molecules on the active sites of the nanoparticles were avoided by using the SFME technique to create the TiO_2 nanoparticles.

Zinc oxide (ZnO) nanoparticles were synthesized from microemulsions in a microchannel reactor system. In order to prevent the production of big particles, microemulsions limit the reactants, which is advantageous for regulated reaction and nucleation. The microemulsions also stop ZnO particles from depositing on the reactor's microchannel walls. According to Wang *et al.*,[37] the production of ZnO nanoparticles was studied using three Zn^{2+} sources: $Zn(NO_3)_2$, $ZnSO_4$, and $ZnCl_2$. The one that performed the best, $Zn(NO_3)_2$,

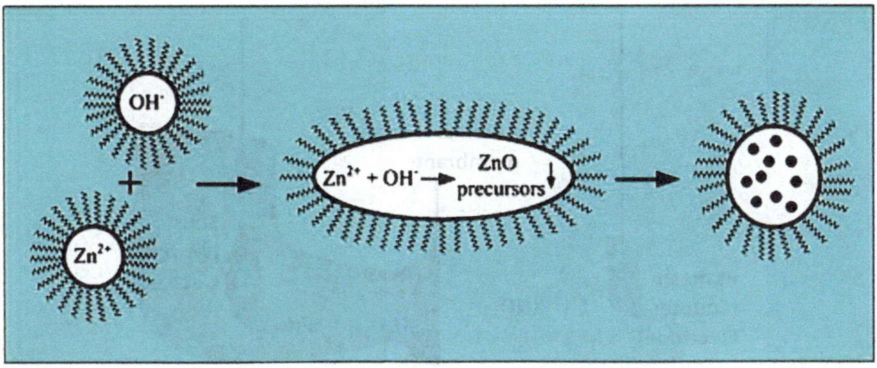

Figure 7.4 Synthesis of ZnO precursor by microemulsions dispersed in an oil phase. Reproduced from ref. 37 with permission from Elsevier, Copyright 2014.

produced ZnO particles with the smallest average grain sizes. On average ZnO nanoparticle particle size, the impacts of Zn^{2+} concentration, reaction temperature, and feed flow rate were examined. ZnO nanoparticles with an average size of 16 nm were produced under ideal circumstances. Figure 7.4 depicts the synthesis of ZnO precursor by microemulsion technique.

7.2.1.4 Electrochemical Method

The synthesis of ceramic thin films and coatings, nanoparticulate materials, and metastable phases can now be produced easily and affordably using emerging electro-synthetic processes such as electro-generation of base, anodic oxidation, and AC (alternating current) synthesis. An electrochemical synthesis is achieved by passing an electric current between two or more electrodes separated by an electrolyte. By definition, the synthesis takes place at the electrode–electrolyte interface. The electric double layer, has an extremely large potential gradient of 105 V cm^{-1}, where electrochemical synthesis takes place near to the electrode. The product is deposited as a thin film or coating on the electrode. Moreover, a solid–liquid interface makes it easier for conformal coatings to grow on substrates of any shape, particularly when a counter electrode with the suitable shape is used to produce uniform polarization. Electrochemical synthesis is a low-temperature technique limited by the boiling point of the electrolyte. This can be raised by using molten salt electrolytes. Controlling the current that flows through the cell allows for the exercise of kinetic control, whereas selecting the applied cell potential allows for thermodynamic control. An oxidation or reduction reaction is the basis of an electrochemical synthesis. The oxidizing or reducing power can be continually changed and appropriately chosen by fine-tuning the applied cell potential.[38]

Many groups have reported on the electrochemical synthesis of precursors to polycrystalline powders of ZrO_2, TiO_2, and $ZrTiO_4$ and coatings of ZrO_2, Al_2O_3, Al_2O_3–ZrO_2, and Al_2O_3–Cr_2O_3. The respective mixed-metal nitrate/

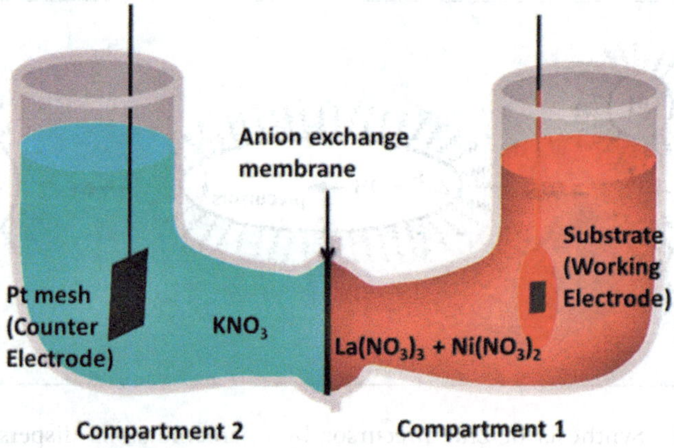

Figure 7.5 Schematic of the electrochemical cell. Reproduced from ref. 38 with permission from Elsevier, Copyright 2019.

chloride solutions were statically reduced in the presence of hydrogen peroxide at different current densities ranging from 10 to 100 mA cm^{-2} to produce these syntheses. Working electrodes made of graphite or Ti were employed. The OH ions produced during these syntheses cause the hydrolyzed compounds of Zr or Ti to form colloidal particles on the cathode, which, when heated, turn into the corresponding metal oxides. This technique is excellent for creating fine powders and adherent coatings[39] (see Figure 7.5).

Similar to this, base electro-generation has been used to create precursors for oxides including PbO_2, Cr_2O_3, Nb_2O_5, ZnO, and WO_3. By altering specific variables, including electrolyte concentration, current density, deposition time, electrolyte pH, and temperature, the oxides can be produced in all of these syntheses as adherent coatings that are only a few micrometres thick.

7.2.1.5 Thermal Decomposition Method

Thermal decomposition is an innovative method to synthesize stable monodispersed nanoparticles product. Compared to traditional methodologies, this study area is rapidly developing, clearer, and more cost-effective. Moreover, it is among the simplest and most practical ways to synthesis monodispersed metal nanoparticles. Also, it provides a solution to the biggest problem in nanotechnology research: attaining a controlled nanometric size and shape. Additional elements to be considered for controlled nanometric size include synthesis duration, temperature, reactant concentrations, stabilizers, capping agents (surfactants), and surfactant types. According to Palacios-Hernández *et al.* and Kino *et al.*, the solventless method of thermal decomposition is a simple, moderate process that does not need any raw materials.[40,41] Similar to this, Tran *et al.* claimed that thermal decomposition allowed for the production of vast numbers of nanoparticles at a single time as opposed to biological methods, which only produced small numbers.

Tran *et al.* also highlighted the importance of the precursor injection technique, in which the precursor was injected into a hot surfactant solution.[42] The goal was to easily create nanoparticles with small diameters that are identical or comparable and have narrow size distributions. The term "homogeneous" could also be used to describe these monodispersed nanoparticles.

The temperature and pressure at which thermal decomposition occurs depend on several variables, including the type of the metallic ion and the strength of reactivity with the ligands in coordination compounds. Coordination chemicals can thus be either kinetically stable or thermodynamically stable. For thermal decomposition, no specific stabilizing agent (stabilizer) is employed. Rao *et al.* discovered that capping agents, such as carboxylic acids and alkyl amines, affected how monodispersed nanoparticles produced during thermal breakdown were formed.[43] Achieving monodispersed nanoparticles is impacted by the total effect. During thermal decomposition, the decreased samples are injected into the surfactants at a specific temperature, and after a period of time in an inert environment, the sample's colour changes to black. Colloidal nanoparticles are indicated by a black solution. Polymers and the functional groups of amines, carboxylic acids, hydroxyls, and thiols are effective capping agents for stabilizer thermal decomposition. Figure 7.6 illustrates the thermal decomposition method.[44]

The thermal decomposition technique has been used to create a wide range of materials, including metal oxides, carbides, nitride, and sulphides. Development of inorganic or organometallic compounds for the synthesis of nanoparticles is currently a prominent area of research. According to study that has been done by Prakash *et al.*, a one-pot thermal decomposition technique was used to successfully synthesize a $NaGdF_4$:Yb:Er/Ag/TiO_2 nanocomposite photocatalyst.[45] By using X-ray diffraction and transmission electron

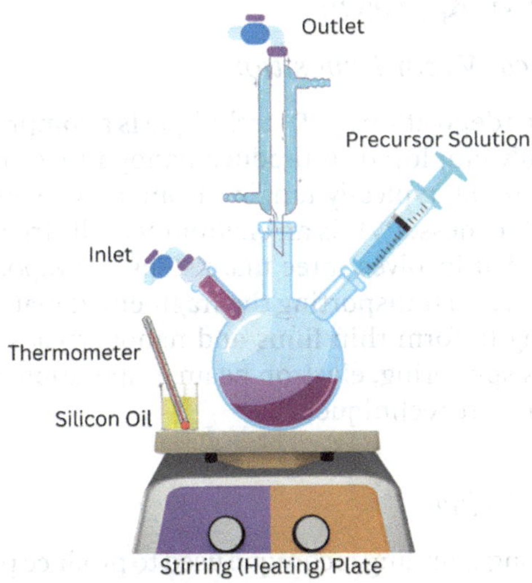

Figure 7.6 Thermal decomposition method.

microscopy analysis, the composite structure, morphology, and elemental mapping of the produced $NaGdF_4$:Yb:Er/Ag/TiO_2 nanocomposite were characterized. Upon conversion photoluminescence measurements at 980 nm were used to highlight the energy transfer between $NaGdF_4$:Yb:Er, Ag, and TiO_2. The visible and NIR light absorption properties of the $NaGdF_4$:Yb:Er/Ag/TiO_2 nanocomposite were improved by the Ag and $NaGdF_4$ nanoparticles. Rhodamine B degradation was used to analyze the NIR and UV-visible light induced photocatalytic research of the $NaGdF_4$:Yb:Er/Ag/TiO_2 composite. The photocatalytic activity under NIR irradiation was significantly impacted by the energy transfer between $NaGdF_4$:Yb:Er, Ag, and TiO_2. The photocatalytic breakdown of rhodamine B is caused by oxidative species that the catalysts produced when exposed to NIR radiation.[45]

According to a study undertaken by Hajiashrafi and Kazemi,[46] ZnO nanoparticles were synthesized by a simple and repeatable method with thermal decomposition of a zinc-based metal organic framework (Zn-MOF). Zn-MOF thermal decomposition is one of the new routes for forming ZnO nanoparticles. Due to their distinct characteristics and current developments in applications, metal organic framework (MOF), also known as porous coordination polymer, has received a great deal of attention from scientists in recent years. MOF is made up of an organic–inorganic crystalline hybrid three-dimensional (3D) network. The coordination of polydendate ligands as organic sectors with metal ions (or metal clusters) as inorganic sectors results in the self-assembly of metal organic frameworks. Due to its substantial surface area, high pore volume, and customizable pore and shape, MOF has garnered a lot of attention. Being one of the most widely used metal–organic coordination polymers, zinc-based metal organic framework is becoming more and more common.

7.2.2 Top-down Approach

7.2.2.1 Physical Vapor Deposition

The physical vapor deposition (PVD) technique is a comprehensive set of procedures commonly employed to produce nanoparticles (NPs) and to apply thin layers of material, typically ranging from a few nanometers to several micrometres in thickness. PVD is an environmentally friendly vacuum deposition technique that involves three main steps: (1) vaporizing the material from a solid source, (2) transporting the fragmented material, and (3) nucleating and growing to form thin films and nanoparticles. The PVD method typically employs sputtering, electron beam evaporation, pulsed laser deposition, and vacuum arc techniques.[47]

7.2.2.2 Melt Mixing

Melt mixing is a long-standing technique used to produce polymeric nanoparticles. Due to its environmentally friendly nature, this mechanical technique is predominantly employed in contemporary industrial applications.

The primary purpose of these approaches is to mechanically blend the polymers with certain nano-fillers.[48] In 2012, Jeziórska *et al.* fabricated nano-composites of low-density polyethylene/spherical silica (LDPE/SGS) using a twin-screw co-rotating extruder using a melt mixing method. The resulting composites exhibited enhanced mechanical strength.[49]

7.2.2.3 *Laser Ablation*

The process of evaporating particles from the solid surface can be achieved with this LA approach. To achieve this procedure, this technology employs a high-power laser beam, either in a continuous manner or in a pulsed manner. This approach will impart polymeric materials at a micro or nano scale.[50] In 2001, Naessens *et al.* conducted a study in which they utilized the LA approach to create an arrangement of micro lenses. They achieved this by scanning the surface of a polycarbonate polymer to achieve a lens form that had the ideal focal distance and diameter.[51]

7.2.2.4 *High Energy Ball Milling*

The high energy ball milling (HEBM) process frequently requires operating at elevated temperature and pressure conditions. There is a potential for the temperature to exceed 1000 °C and the pressure to reach several gigapascals (GPa). The origin of this strategy may be traced back to 1970 when it was initially devised by John Benjamin. During this process, the kinetic energy of the balls is used to break the chemical bonds of the bulk material, leading to the crushing of the material and the synthesis of smaller nanoparticles. This also leads to the creation of new surfaces.[52] In 2017, S. Ghorban Hosseini and E. Ayoman produced haematite (α-Fe_2O_3) nanoparticles using HEBM in a dry environment. The nanoparticles had a size range of 23–20 nm and were synthesized for various milling durations of 40–60 hours.[53]

7.3 Modification in Metal Oxides for Pollutant Removal

TiO_2 is frequently used to stimulate the breakdown of organic molecules under UV-Vis irradiation, which aids in the photodegradation of numerous chemical contaminants found in wastewater. Pure TiO_2 is more prone to excitation when exposed to UV light because of its broad bandgap, and the rate of recombination of photogenerated species electron–hole (e^-/h^+) pairs is considerable. Despite this, the photocatalytic efficiency of pure TiO_2 is still rather low. A model photocatalyst should have a bandgap that allows for visible light excitation. Such nanostructured photocatalysts should have favourable valence band (VB) and conduction band (CB) positions for redox reactions, as well as good charge separation efficiency, charge carrier, and transport. Since the UV region only makes up a very minor portion of the solar spectrum, visible-light active nano photocatalysts are chosen in this situation. As a result, one way to make TiO_2 absorb visible light is to alter its inner band structure by substituting oxygen in the TiO_2 lattice to reduce the optical energy bandgap.[54]

7.3.1 Doping of Metal-to-metal Oxides

Regarding this matter, several methods (structural and chemical modifications) have been developed to improve the photocatalytic activity of TiO_2 and get around some of the issues described above, particularly bandgap energy and recombination rate.[55] The ability of TiO_2 to capture visible light is thought to be increased by doping it with metals and/or non-metals, which prevents electron–hole recombination and boosts photocatalytic effectiveness. TiO_2 can have its ability to absorb visible light or its UV reactivity increased by doping it with metal or non-metal ions. As a result of reducing the photocatalysis capability in conjunction with thermal instability, metallic elements can cause a sufficient bandgap shift and promote recombination. Also, when noble metal ions are deposited on a TiO_2 surface, they serve as light collectors and improve the catalyst's activity in the visible light spectrum. Thus, it is proven by previous research that doping of TiO_2 with various elements increases its photocatalytic activity due to the formation of new energy levels near the conduction band. Titanium dioxide photocatalysis is a heterogeneous process in which the catalyst's surface has a significant impact. TiO_2's structural characteristics are affected by the synthesis process, the doping process, and the dopants.[56] In Table 7.2 the photocatalytic performance of various noble metal doped TiO_2 composites is discussed.[57]

7.3.2 Doping of Non-metal-to-metal Oxides

According to current knowledge, the most essential method for improving the photoactivity of undoped TiO_2 is to dope it with non-metal components such as nitrogen, carbon, fluorine, chlorine, or sulphur. Because of its size resemblance to an oxygen atom, high ionization, and high electronegativity energies, nitrogen is the non-metal element that is most effective at doping. Several studies have shown that nitrogen doping of TiO_2 can increase its photocatalytic activity both under UV and visible light. The choice of foreign nitrogen ion sources and the chemical states of the dopants are also important factors in the high activity of the N-TiO_2 catalyst. In order to create N-TiO_2 photocatalysts with excellent photocatalytic characteristics, the choice of nitrogen ion sources is essential. N-TiO_2 photocatalyst has been created utilizing a variety of nitrogen sources, including ammonia, nitrogen dioxide, thiourea, urea, triethylamine, and hydrazine, just like in the earlier experiments. Several studies have suggested that adding urea, a nitrogen precursor, can boost the photocatalytic efficiency of TiO_2. Moreover, Cong *et al.* demonstrated that the anatase crystalline phase can be enhanced by the urea N-TiO_2 catalyst in comparison to undoped TiO_2.[58]

Table 7.2 Photocatalytic performance of noble metal-doped TiO_2. Adapted from ref. 57 with permission from MDPI, Copyright 2020.

Material	Preparation procedure	Reaction conditions	Elimination efficiency
Mesoporous TiO_2 doped with Pd and Au	Sol–gel	Simulated wastewater with 8.71 mg L^{-1}, Total Organic Carbon (TOC), 15 W UV lamp	93% (Au/TiO_2)
Pd-, Pt-, Ag- or Cu-doped TiO_2 (P-25) 0.5% Pt/P-25 10% Ag/P-25	US treatment and UV light in water-propanolImpregnation from polyol solutions under reflux	2,2′,4,4′-Tetrabromodiphenyl ether, 300 W UV lamp	100% on 5%Pd-TiO_2
1%Pt/TiO_2/SiO_2 (a) 1%Pd/TiO_2/SiO_2 (b) 1%Ag/TiO_2/SiO_2	Photo-deposition in acetic acid suspensions of TiO_2/SiO_2 containing $PdCl_2$ or $H_2PtCl_6{\cdot}6H_2O$ under UV (8 W lamp) for 15 h	Brilliant red (K-2G), cationic blue (CBX), 300 W UV lamp	K-2G 280% rate increase on (a), 30% rate increase on (b)
0.5% and 1% Ag and Pt/P-25	Photo-deposition in methanol (Ag) and oxalic acid solution (Pt) under UV-C (4 W lamp) for 5 h	Oxalic acid, 9 W UV-A lamps	70–85% on doped samples, 40% on P-25
0.1% Ag and/or 0.1% Pd and Pt/P-25	Radiolysis	Toluene phenol, 63 mW UV-LED, or Vis-LED	50–97% or 60–100% depending on the doping
TiO_2 and Pd/TiO_2	Sol–gel method (Ti isopropoxide and Pd nitrate)	Methylene blue, methyl orange, 100 W UV lamp	83.4 and 75.3% mineralization after 180 min

Sol–gel, hydrothermal, electrochemical anodic oxidation, vapor deposition, and calcination of $Ti(OH)_4$ under nitrogen atmospheres were some of the techniques used to prepare the N-TiO_2 photocatalyst. Since different methods for integrating nitrogen into the TiO_2 lattice might result in a varied placement of the nitrogen atom in the structure and with varying photoactivities, the preparation process played a crucial influence in defining the final attributes of the N-TiO_2. Because it is regarded as convenient, easy to use, and ecologically safe, the hydrothermal approach was used to create N-TiO_2. It is also one of the best ways to produce high-quality TiO_2 with characteristics like high anatase crystallinity, high specific surface area, and narrow particle size dispersion. Moreover, using this technique, TiO_2 crystallites can be created without being subjected to a high-temperature calcination process.

For the photocatalytic degradation of ciprofloxacin, nitrogen was doped into TiO_2, according to earlier studies. Using urea (NH_2CONH_2) as a nitrogen source, the hydrothermal method was used to create the N-TiO_2 photocatalyst. This study examined the impact of different nitrogen doping concentrations on the surface morphology, crystalline phase structure, optical and textural properties, chemical state, and photocatalytic activity of N-TiO_2. Focused ion beam–field emission scanning electron microscopy (FIB/FESEM) was utilized to capture the surface morphologies. The Brunauer–Emmett–Teller (BET) model was used to calculate surface area values. Using the Barrett–Joyner–Halenda (BJH) hypothesis as it relates to the desorption of the nitrogen isotherm, the pore size distributions of the catalysts were calculated. Wide-angle X-ray scattering was used to examine the crystallite and phase structure of all N-TiO_2 photocatalysts (WAXS). Using UV-VIS diffusion reflectance spectroscopy (UV-VIS-DRS) and the Kubelka–Munk function, the bandgap energy was ascertained. X-ray photoelectron spectroscopy was used to examine the chemical states (XPS). The photo-illumination of ciprofloxacin under UV light was used to assess the photocatalytic activity of the synthesized N-TiO_2.[59]

7.3.3 Doping of Polymers to Metal Oxides

Due to their low cost, superior film forming ability, good mechanical and anti-compaction capabilities, strong chemical and thermal stabilities, and exceptional acidic and alkaline resistance, polymeric membranes have found use in ultrafiltration processes in many industrial domains. Membranes also benefit from a small footprint, improved separation effectiveness, and simple maintenance.[60] However, due to severe membrane fouling and a decrease in water flux, their hydrophobic nature has made it difficult for them to be applied in the treatment of water and wastewater.[61] Membrane fouling necessitates careful attention since it can lead to increased energy use, expensive membrane cleanup, and frequent membrane replacement.[62] This prevents their use in wastewater with a high pollution load, limiting their use to water polishing in the treatment train. Multifunctional mixed matrix membranes (MMM) have been the focus of recent developments in membrane technology. These membranes are created by incorporating inorganic nanoparticles either *in situ* or *ex situ*.[63,64] A fascinating area of study in water and wastewater treatment is hybrid technologies that combine photocatalysis with membrane technology.[60] The advantages of connecting the processes allow for the membrane-based retention of the photocatalyst in the reaction medium, control of particle retention duration in the reactor, and implementation of a continuous water treatment train.[65]

A readily available engineering substance called polysulfone (PSf) (see Figure 7.7) is renowned for its exceptional mechanical, thermal, chemical, and environmental endurance. There has been a lot of interest in changing the surface of the PSf membrane using physical or chemical means in order to increase its hydrophilicity. The physical and chemical characteristics of PSf membranes have been somewhat improved through the application of modification techniques such UV radiation, blending with hydrophilic materials,

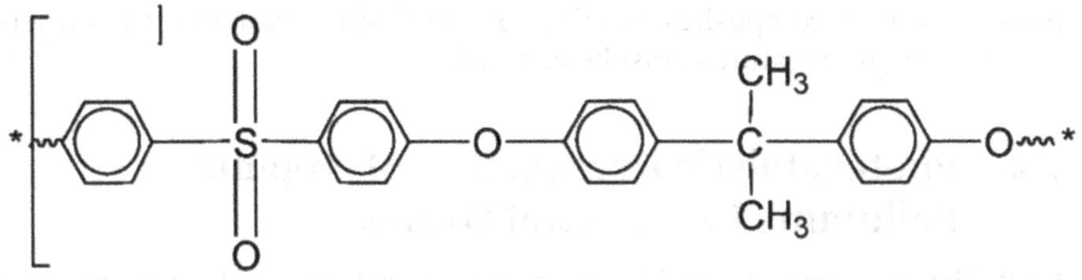

Figure 7.7 Repeat unit, chemical structure of polysulphone. Reproduced from ref. 66 with permission from Elsevier, Copyright 2018.

graft polymerization, dip and spin coating, and production of thin films on the membrane surface. One of these methods, the insertion of TiO_2 nanoparticles into membranes, has given rise to a brand-new integrated water treatment method known as the photocatalytic membrane reactor or membrane photocatalytic reactor. Such reactors can take the shape of a slurry photocatalytic reactor followed by a membrane filtration unit, a polymeric membrane submerged in a slurry photocatalytic reactor, a membrane placed inside a photo-reactor whose internal walls are coated with a photocatalyst, or a photocatalytic membrane made of pure TiO_2 or a TiO_2 composite membrane. The ideal solution should include the advantages of photocatalysis, membrane filtration, organic degradation, and TiO_2's antibacterial capabilities in a single unit. Since intrinsic TiO_2 has a significant band gap (E_g = 3.2 eV), it can only be activated by UV light.[66]

Co-doping is said to have narrowed the band gap and increased TiO_2's photocatalytic efficiency. The optical characteristics of PSf membranes will probably change as a result of the co-doped TiO_2 incorporation, which may make it possible for the co-doped TiO_2 to be activated photocatalytically under less intense visible light illumination. The stability of polymeric membranes under high energy UV irradiation is currently not well characterized, especially for prolonged exposure times.[67,68] According to reports, adding TiO_2 to PSf membranes improves their hydrophilicity, mechanical strength, thermal stability, permeability, and rejection capacity.[69] The increased pure water flux and humic acid permeability of UV-irradiated TiO_2 coated PSf membrane were reported by Mohamad *et al.*[70] In a different investigation, it was discovered that TiO_2 may reduce the fouling of PSf hollow fiber membranes with humic acid as a result of concentration polarization, cake layer development, and absorption.[71] Under the influence of sunshine, Cr(VI) was reduced to Cr(III) using PSf/TiO_2 composite membranes. The study showed that hybrid membranes may be recycled six times without any pre-treatment and without losing any photocatalytic activity.[72] These findings emphasize the importance of TiO_2 in enhancing the chemical and physical characteristics of the PSf membrane. Although photocatalytic membranes have been created, more investigation is still required to fully comprehend their structure and mode of operation. To produce effective, inexpensive, and energy-efficient membrane nanocomposites suitable for the desired wastewater treatment application, their features must be clarified.

7.4 Photocatalytic Degradation of Organic Pollutants Using Metal Oxides

Pesticides are widely utilized in agriculture, gardening, and a wide range of other domestic applications. Chlorpyrifos is an organophosphorus pesticide that has been used extensively throughout the world to control insect pests in soil, ornamental plants, fruits, and vegetables.[73] Organic pollutants are removed during the photo-catalytic oxidation process in the presence of semiconductor photocatalysts (such as TiO_2 and ZnO), an energetic light source, and an oxidizing agent, such as oxygen or air.[74] The electron in the conduction band decreases the oxygen adsorbed on the photocatalysts, while the positive hole either directly oxidizes the pollutant or oxidizes water to produce hydroxyl radicals ($^{\bullet}OH$ radical) (TiO_2 or ZnO).[75]

7.4.1 Photocatalytic Degradation of Pesticides Using Metal Oxides

In the degradation of organic pollutants, the hydroxyl radical is produced by the oxidation of adsorbed water, the principal oxidant, and the presence of oxygen can prevent the recombination of an electron–hole pair. Organic substances like insecticides, chlorinated aromatics, and nitro phenols are attacked by the hydroxyl radical. In recent years, TiO_2 and ZnO have been employed as efficient, affordable, and non-toxic semiconductor photocatalysts for the oxidation of a variety of organic compounds.[76] Titanium dioxide (TiO_2) has proven to be an effective catalyst due to its stability and non-toxicity, and its behavior is quite well-documented in literature.[77] The photocatalytic action of other semiconductors, such as zinc oxide (ZnO), is less well understood. Because of its direct band gap, high excitation binding energy, and piezoelectric characteristics, ZnO is a very intriguing wide-band-gap semiconductor material. This semiconductor material has a band gap of about 3.2 eV, which translates to a radiation wavelength of about 390 nm. To excite the electrons in the valence band to the conduction band, a UV light with a wavelength shorter than 380 nm is required.

Using semiconductor oxide catalysts, namely zinc oxide (ZnO) and titanium dioxide, the degradation of chlorpyrifos in water was studied in the current work (TiO_2). It was also investigated how different parameters, including pH, sodium bicarbonate salt, catalyst concentrations, irradiation time, and type of catalyst, affected the results. The catalyst's optimum concentration, according to the results, was 0.15 g L^{-1}. Using the same photocatalytic reaction conditions, it was discovered that TiO_2 is a superior catalyst to ZnO. At pH 9, the greatest elimination effectiveness was attained. The photodegradation efficiency of pesticides is thought to rise with longer illumination times, according to study results. The photodegradation efficiency of chlorpyrifos was reported to be 80% and 90% for UV/ZnO and UV/TiO_2, respectively. Compared to when the salt was not present, photodegradation happened more slowly in the presence of sodium bicarbonate. Moreover, distilled water has a higher photocatalytic degradation efficiency than regular water.[78]

7.4.2 Photodegradation of Dyes Using Metal Oxides

The dyes used in paper, textile, paint, food, plastic, and beverage industries is one of the major contributors to water pollution. These industries utilize multiple types of dyes[79] and 10%–20% of the dyes from these industries are released into water bodies without any treatment. The dyes released into water are difficult to degrade under natural conditions. Hence, several approaches have been developed to remove dyes from polluted water,[80] such as physical and chemical processes like the chemical precipitation method, coagulation and electrocoagulation *etc.*, Some of the advance methods used for the removal of dyes include adsorption,[81] photo-catalysis, electrolysis[82] *etc.* Each method has its own advantages and disadvantages. Furthermore, photocatalysis is an advanced oxidative process that efficiently and rapidly oxidises a wide range of organic contaminants. It is also a simple and low-cost method for the removal of both cationic and anionic dyes. Electrolysis works on the redox reaction, by transferring electrons with toxic aliphatic and aromatic organic pollutants and immediately forming decolorized intermediates of pollutants as byproducts.

Development of multi-application material is much more advantageous, compared to materials intended for specific applications.[83] Among various nanometal oxides, TiO_2 and its composites was utilized as photocatalyst and adsorbent for removal of many dyes. Synthesized composite Ti/TiO_2/RuO_2 was used to remove procion blue of the textile wastewater using photocatalysis and electrolysis methods.[84] UV/nano-TiO_2 having a homogeneous structure was used for the removal of methylene blue dye by photocatalysis.[85] The TiO_2–SiO_2 composite was prepared using commercial TiO_2 which in the presence of UV irradiation degraded the methylene blue dye.[86] Using commercial TiO_2 prepared TiO_2@yeast–carbon microspheres are used for the removal of cationic dye methylene blue and anionic dye congo red by the adsorption process.[87] TiO_2–Fe_3O_4-bentonite magnetic nanocomposites were used for the removal of azo dye methylene blue using photocatalysis.[88]

7.4.3 Photodegradation of Drugs Using Metal Oxides

Pharmaceuticals are one of the most significant emerging environmental contaminants, and other micropollutants that were previously undetected are increasingly affecting drinking water sources and water quality. Modern technologies are enabling the increasing development and production of pharmaceuticals that serve to maintain quality of life. Cancer is the second-leading cause of death worldwide, therefore among the various types of active pharmaceutical substances, anticancer medications have attracted greater attention and need to be monitored in the environment as their production and consumption rise.[89] Their relatively good solubility, polarity, increased mobility, and persistence make it easier for them to enter the environment through healthcare facilities, improper disposal, domestic use, and ultimately wastewater treatment facilities that are not made to filter out small molecules like pharmaceuticals.[90] Many anticancer medications pose a direct threat to aquatic and terrestrial organisms due to their potential carcinogenic, mutagenic, genotoxic, and ecotoxic effects; however, the development of more sensitive analytical techniques over many years allows for their detection at true g/ng levels, which should not be disregarded.[91] Several biotic and abiotic degradation processes have an impact on the stability of the drug after release into the environment, which is typically excreted in a mixture with metabolites. This results in potentially more harmful degradation and transformation products.

For the successful protection of water sources and the possibility of their reuse, advanced oxidation processes such as photocatalysis are of great interest due to their wide application in the degradation of various organic molecules through the generation of non-selective and highly oxidative oxygen species.[92,93] Due to its favourable characteristics, including chemical and thermal stability, low cost, convenience of use, and strong photocatalytic efficiency in anatase crystal modification, TiO_2 continues to be the most widely used semiconductor material for photocatalytic reactions. Natural sunlight, or UVA irradiance, can activate TiO_2 at a photon energy sufficient to bridge the band gap between the valence and conduction bands (3.2 eV), which considerably reduces the energy use and toxicity of commonly used UVC lamps.[94,95] Sol–gel immobilization of TiO_2 on a solid support is one method for resolving issues with the separation of powder and water in suspension form and the reuse of the catalyst with the potential for its dispersion in the environment, which ultimately enables easier and less expensive commercialization of the developed system.[96] TiO_2 that has been exposed to light energy creates electron–hole pairs that decrease oxygen or oxidize water to create reactive species that oversee pollutant degradation. Scavengers are organic and inorganic compounds that can be introduced to an aqueous solution to disrupt the reactive species participating in the degradation in order to assess the mechanism of heterogeneous photocatalysis. As a result, alcohols like isopropanol and *tert*-butanol are used to verify the activity of hydroxyl radicals ($^{\bullet}OH$), *p*-benzoquinone scavenges superoxide ion radicals, azide reacts with singlet oxygen, and formic acid is utilized to determine the contribution of the holes in the valence band.[59,97]

The drugs such as imatinib and crizotinib which are used to treat leukaemia and lung cancer, respectively are thought to be very persistent, potentially non-biodegradable substances that are prevalent in the aquatic environment.[98] These two pharmaceuticals, along with other anticancer medications, cannot be removed from water using traditional water treatment techniques, so photocatalysis was investigated to study the degradation kinetics under the mutual influence of contaminant concentration at an environmentally relevant pH. Response surface modelling (RSM) and design-expert software were used to identify the ideal process parameters. The photocatalyst was TiO_2 immobilized on a glass fibre mesh. To better understand the reaction mechanism, scavenger research was conducted. In addition, certain organic and inorganic chemicals that are naturally present in water matrices might have a positive or negative impact on the photocatalytic efficacy in drug removal. Consequently, the effect of humic acids and bicarbonate, phosphate, nitrate, and chloride ions in an aqueous solution of medicines was evaluated in recent work. The tentative structures of the degradation products generated under ideal photocatalytic conditions were clarified using high-resolution mass spectrometry (HPLC-QqTOF). Lastly, employing sea bacteria called *Vibrio fischeri* as test organisms, the acute toxicity of medicines and their combination with breakdown products was investigated.[99]

7.5 Conclusion

There are several previously known water purification methods such as coagulation, ion exchange, adsorption processes, ozonation, electrochemical techniques, *etc.* The traditional treatment technologies were proved to be ineffective for fully degrading recalcitrant pollutants in wastewater. However, photocatalysis appears to be an interesting approach for water purification, with the possibility of using UV/visible light (sustainable as solar light can be used after doping because of the reduction in band gap energy). Doping these catalysts with metals and/or non-metals have been proved to be an efficient way to boost its photocatalytic efficiency because it prevents electron–hole recombination and increases its ability to harvest visible light. The synthesis of metal oxides for pollutant removal has been carried out using a variety of chemical and physical techniques which can be further divided into bottom-up (chemically) and top-down (physically). The top-down techniques involve physical vapor deposition, melt mixing, laser ablation, and high energy ball milling (HEBM) *etc.* The bottom-up techniques include sol–gel, hydrothermal, microemulsion, electrochemical, thermal decomposition, co-precipitation, and biological *etc.* Each of these synthesis techniques have their unique features and benefits such as the micro-emulsion method is a modern technique to create nanoparticles. The metal oxide (like TiO_2/ZnO) nanoparticles produced through this method showed excellent degradation efficiency towards pollutant. The hydrothermal technique is easy to use, convenient, and ecologically safe. The electrochemical technique is useful for creating fine powders and adherent coatings. Furthermore, the thermal decomposition method is found to be better suited for rapidly developing, cost-effective monodispersed nanoparticles. Amongst the top-down approaches, the physical vapor deposition technique can be used for preparing thin layer materials. For producing polymeric nanoparticles, the melt mixing technique is generally helpful. Furthermore, due to the doping of metals, non-metals, and other moieties to metal oxides are found to be effective for the degradation of organic pollutants. Moreover, the photodegradation employing doped metal oxides for the breakdown of various organic contaminants such as pesticides, drugs, and dyes was found to be better in visible light too. In conclusion, this chapter consolidates the significance of photocatalytic materials in combating water pollution.

Acknowledgments

The authors appreciate the Research Supporting Project (304/PKIMIA/6315580) by Universiti Sains Malaysia.

References

1. P. Rajasulochana and P. Vijay, Comparison on efficiency of various techniques in treatment of waste and sewage water – A comprehensive review, *Resour.-Effic. Technol.*, 2016, **2**, 175.
2. P. A. Owusu, S. Asumadu-Sarkodie and P. Ameyo, A review of Ghana's water resource management and the future prospect, *Cogent Eng.*, 2016, **3**, 1164275.
3. L. Ma, X. Dong, M. Chen, L. Zhu, C. Wang and F. Yang, *et al.*, Fabrication and Water Treatment Application of Carbon Nanotubes (CNTs)-Based Composite Membranes: A Review, *Membranes*, 2017, **7**, 16.
4. Y. Zhang, I. Karatchevtseva, M. Bhadbhade, T. T. Tran, I. Aharonovich and D. J. Fanna, *et al.*, Solvothermal synthesis of uranium(VI) phases with aromatic carboxylate ligands: A dinuclear complex with 4-hydroxybenzoic acid and a 3D framework with terephthalic acid, *J. Solid State Chem.*, 2016, **234**, 22.
5. X. Tang, H. Zheng, H. Teng, Y. Sun, J. Guo and W. Xie, *et al.*, Chemical coagulation process for the removal of heavy metals from water: a review, *Desalin. Water Treat.*, 2016, **57**, 1733.
6. J. A. Gallego-Urrea, J. Hammes, G. Cornelis and M. Hassellöv, Coagulation and sedimentation of gold nanoparticles and illite in model natural waters: Influence of initial particle concentration, *NanoImpact*, 2016, **3–4**, 67.
7. D. Qadir, H. Mukhtar and L. K. Keong, Synthesis and Characterization of Polyethersulfone/Carbon Molecular Sieve Based Mixed Matrix Membranes for Water Treatment Applications, *Procedia Eng.*, 2016, **148**, 588.
8. M. L. Pype, M. G. Lawrence, J. Keller and W. Gernjak, Reverse osmosis integrity monitoring in water reuse: The challenge to verify virus removal – A review, *Water Res.*, 2016, **98**, 384.
9. J. M. Dickhout, J. Moreno, P. M. Biesheuvel, L. Boels, R. G. H. Lammertink and W. M. de Vos, Produced water treatment by membranes: A review from a colloidal perspective, *J. Colloid Interface Sci.*, 2017, **487**, 523.
10. A. Subramani and J. G. Jacangelo, Emerging desalination technologies for water treatment: A critical review, *Water Res.*, 2015, **75**, 164.

11. I. Muñoz, A. Rodríguez, R. Rosal and A. R. Fernández-Alba, Life Cycle Assessment of urban wastewater reuse with ozonation as tertiary treatment: A focus on toxicity-related impacts, *Sci. Total Environ.*, 2009, **407**, 1245.
12. L. Quartaroli, B. H. Cardoso, G. de Paula Ribeiro and G. H. Da Silva, Wastewater chlorination for reuse, an alternative for small communities, *Water Environ. Res.*, 2018, **90**, 2100.
13. A. S. Adeleye, J. R. Conway, K. Garner, Y. Huang, Y. Su and A. A. Keller, Engineered nanomaterials for water treatment and remediation: Costs, benefits, and applicability, *Chem. Eng. J.*, 2016, **286**, 640.
14. F. C. Moreira, J. Soler, M. F. Alpendurada, R. A. R. Boaventura, E. Brillas and V. J. P. Vilar, Tertiary treatment of a municipal wastewater toward pharmaceuticals removal by chemical and electrochemical advanced oxidation processes, *Water Res.*, 2016, **105**, 251.
15. N. A. Ramos-Delgado, M. Á. Gracia-Pinilla, R. V. Mangalaraja, K. O'Shea and D. D. Dionysiou, Industrial synthesis and characterization of nanophotocatalysts materials: titania, *Nanotechnol. Rev.*, 2016, **5**, 467.
16. B. Srikanth, R. Goutham, R. Badri Narayan, A. Ramprasath, K. P. Gopinath and A. R. Sankaranarayanan, Recent advancements in supporting materials for immobilised photocatalytic applications in waste water treatment, *J. Environ. Manage.*, 2017, **200**, 60.
17. L. V. Bora and R. K. Mewada, Visible/solar light active photocatalysts for organic effluent treatment: Fundamentals, mechanisms and parametric review, *Renewable Sustainable Energy Rev.*, 2017, **76**, 1393.
18. J. Kou, C. Lu, J. Wang, Y. Chen, Z. Xu and R. S. Varma, Selectivity Enhancement in Heterogeneous Photocatalytic Transformations, *Chem. Rev.*, 2017, **117**, 1445.
19. M. R. D. Khaki, M. S. Shafeeyan, A. A. A. Raman and W. M. A. W. Daud, Application of doped photocatalysts for organic pollutant degradation - A review, *J. Environ. Manage.*, 2017, **198**, 78.
20. S. K. Afridi, K. Umar, T. Parveen and M. R. Razali, in *Nanocomposites-Advanced Materials for Energy and Environmental Aspects*, ed. M. E. Khan, J. Aslam, and C. Verma, Woodhead Publishing, Cambridge, 1st edn, 2023, ch. 16, p. 321.
21. M. Pirhashemi, A. Habibi-Yangjeh and S. Rahim Pouran, Review on the criteria anticipated for the fabrication of highly efficient ZnO-based visible-light-driven photocatalysts, *J. Ind. Eng. Chem.*, 2018, **62**, 1.
22. K. Umar, M. N. M. Ibrahim, A. Ahmad and M. Rafatullah, Synthesis of Mn-doped TiO_2 by novel route and photocatalytic mineralization/intermediate studies of organic pollutants, *Res. Chem. Intermed.*, 2019, **45**, 2927.
23. C. B. Anucha, I. Altin, E. Bacaksiz and V. N. Stathopoulos, Titanium dioxide (TiO_2)-based photocatalyst materials activity enhancement for contaminants of emerging concern (CECs) degradation: In the light of modification strategies, *Chem. Eng. J. Adv.*, 2022, **10**, 100262.
24. K. Umar, A. Aris, H. Ahmad, T. Parveen, J. Jaafar and Z. A. Majid, *et al.*, Synthesis of visible light active doped TiO_2 for the degradation of organic

pollutants—methylene blue and glyphosate, *J. Anal. Sci. Technol.*, 2016, **7**, 29.

25. P. K. Dikshit, J. Kumar, A. K. Das, S. Sadhu, S. Sharma and S. Singh, *et al.*, Green Synthesis of Metallic Nanoparticles: Applications and Limitations, *Catalysts*, 2021, **11**, 902.
26. M. G. Motitswe, K. O. Badmus and L. Khotseng, Development of Adsorptive Materials for Selective Removal of Toxic Metals in Wastewater: A Review, *Catalysts*, 2022, **12**, 1057.
27. A. Feinle, M. S. Elsaesser and N. Hüsing, Sol–gel synthesis of monolithic materials with hierarchical porosity, *Chem. Soc. Rev.*, 2016, **45**, 3377.
28. I. Hong, VOCs Degradation Performance of TiO_2 Aerogel Photocatalyst Prepared in SCF Drying, *J. Ind. Eng. Chem.*, 2006, **12**, 918.
29. D. Bokov, A. Turki Jalil, S. Chupradit, W. Suksatan, M. Javed Ansari and I. H. Shewael, *et al.*, Nanomaterial by Sol-Gel Method: Synthesis and Application, *Adv. Mater. Sci. Eng.*, 2021, **2021**, 5102014.
30. S. G. Ullattil and P. Periyat, in *Sol-gel Materials for Energy, Environment and Electronic Applications*, ed. S. C. Pillai, and S. Hehir, Springer, Cham, 1st edn, 2017, ch 9, p. 271.
31. J. N. Hasnidawani, H. N. Azlina, H. Norita, N. N. Bonnia, S. Ratim and E. S. Ali, Synthesis of ZnO Nanostructures Using Sol-Gel Method, *Procedia Chem.*, 2016, **19**, 211.
32. W. Jia, J. Li, Z. Lu, Y. Juan and Y. Jiang, Synthesis of Honeycomb-Like Co_3O_4 Nanosheets with Excellent Supercapacitive Performance by Morphological Controlling Derived from the Alkaline Source Ratio, *Materials*, 2018, **11**, 1560.
33. N. Erdoğan, L. Agartan, J. Park and A. Ozturk, in *Advances in Materials Science for Environmental and Energy Technologies IV*, ed. J. Matyas, T. Ohji, G. Pickrell, W. Wong-Ng, and R. Kanakala, Wiley & Sons, New Jersey, 4th edn, 2015, vol. 253, p. 177.
34. M. A. Malik, M. Y. Wani and M. A. Hashim, Microemulsion method: A novel route to synthesize organic and inorganic nanomaterials: 1st Nano Update, *Arabian J. Chem.*, 2012, **5**, 397.
35. P. Pallavi, K. Harini, S. Alshehri, M. M. Ghoneim, A. Alshlowi and P. Gowtham, *et al.*, From Synthetic Route of Silica Nanoparticles to Theranostic Applications, *Processes*, 2022, **10**, 2595.
36. X. Cui, J. Wang, X. Zhang, Q. Wang, M. Song and J. L. Chai, Preparation of Nano-TiO_2 by Surfactant-Free Microemulsion-Hydrothermal Method and Its Photocatalytic Activity, *Langmuir*, 2019, **35**, 9255.
37. Y. Wang, X. Zhang, A. Wang, X. Li, G. Wang and L. Zhao, Synthesis of ZnO nanoparticles from microemulsions in a flow type microreactor, *Chem. Eng. J.*, 2014, **235**, 191.
38. S. Rehman, A. Shaur, R. Song, T. Lim, J. Hong, S. Park and S. Lee, Nano-fabrication of a high-performance $LaNiO_3$ cathode for solid oxide fuel cells using an electrochemical route, *J. Power Source*, 2019, **429**, 97.
39. G. H. A. Therese and P. V. Kamath, Electrochemical Synthesis of Metal Oxides and Hydroxides, *Chem. Mater.*, 2000, **12**, 1195.

40. T. Palacios-Hernández, G. A. Hirata-Flores, O. E. Contreras-López, M. E. Mendoza-Sánchez, I. Valeriano-Arreola and E. González-Vergara, *et al.*, Synthesis of Cu and Co metal oxide nanoparticles from thermal decomposition of tartrate complexes, *Inorg. Chim. Acta*, 2012, **392**, 277.
41. T. Kino, T. Kuzuya, K. Itoh, K. Sumiyama, T. Wakamatsu and M. Ichidate, Synthesis of Chalcopyrite Nanoparticles via Thermal Decomposition of Metal-Thiolate, *Mater. Trans.*, 2008, **49**, 435.
42. Q. H. Tran, V. Q. Nguyen and A. T. Le, Silver nanoparticles: synthesis, properties, toxicology, applications and perspectives, *Adv. Nat. Sci.: Nanosci. Nanotechnol.*, 2013, **4**, 033001.
43. C. N. R. Rao, P. J. Thomas and G. U. Kulkarni, *Nanocrystals:: Synthesis, Properties and Applications*, Springer, Berlin, Heidelberg, 1st edn, 2007.
44. P. Jain, B. Patidar and J. Bhawsar, Potential of Nanoparticles as a Corrosion Inhibitor: A Review, *J. Bio. Tribo Corros.*, 2020, **6**, 43.
45. N. Prakash, D. Thangaraju, K. Rajan, M. Arivanandhan, Y. Shimura and Y. Hayakawa, UV-visible and near-infrared active NaGdF4: Yb: Er/Ag/TiO_2 nanocomposite for enhanced photocatalytic application, *RSC Adv.*, 2016, **6**, 80655.
46. S. Hajiashrafi and N. M. Kazemi, Preparation and evaluation of ZnO nanoparticles by thermal decomposition of MOF-5, *Heliyon*, 2019, **5**, 02152.
47. X. H. Ma, K. H. Cho and Y. M. Sung, Growth mechanism of vertically aligned SnSe nanosheets via physical vapour deposition, *CrystEngComm*, 2014, **16**, 5080.
48. B. Lin, U. Sundararaj and P. Pötschke, Melt Mixing of Polycarbonate with Multi-Walled Carbon Nanotubes in Miniature Mixers, *Macromol. Mater. Eng.*, 2006, **291**, 227.
49. R. Jeziórska, B. Świerz-Motysia, M. Zielecka, A. Szadkowska and M. Studziński, Structure and mechanical properties of low-density polyethylene/spherical silica nanocomposites prepared by melt mixing: The joint action of silica's size, functionality, and compatibilizer, *J. Appl. Polym. Sci.*, 2012, **125**, 4326.
50. Y. H. Chen and C. S. Yeh, Laser ablation method: use of surfactants to form the dispersed Ag nanoparticles, *Colloids Surf., A*, 2002, **197**, 133.
51. K. Naessens, H. Ottevaere, P. Van Daele and R. Baets, Flexible fabrication of microlenses in polymer layers with excimer laser ablation, *Appl. Surf. Sci.*, 2003, **208**, 159.
52. T. P. Yadav, R. M. Yadav and D. P. Singh, Mechanical Milling: a Top Down Approach for the Synthesis of Nanomaterials and Nanocomposites, *Nanosci. Nanotechnol.*, 2012, **2**, 22.
53. S. G. Hosseini and E. Ayoman, Synthesis of α-Fe_2O_3 nanoparticles by dry high-energy ball-milling method and investigation of their catalytic activity, *J. Therm. Anal. Calorim.*, 2017, **128**, 915.
54. V. Etacheri, C. Di Valentin, J. Schneider, D. Bahnemann and S. C. Pillai, Visible-light activation of TiO_2 photocatalysts: Advances in theory and experiments, *J. Photochem. Photobiol., C*, 2015, **25**, 1.

55. K. Umar, T. Parveen, M. A. Khan, M. I. Mohamad Nasir, A. Ahmad and M. Rafatullah, Degradation of organic pollutants using metal-doped TiO_2 photocatalysts under visible light: a comparative study, *Desalin. Water Treat.*, 2019, **161**, 275.
56. A. Khlyustova, N. Sirotkin, T. Kusova, A. Kraev, V. Titov and A. Agafonov, Doped TiO_2 : the effect of doping elements on photocatalytic activity, *Mater. Adv.*, 2020, **1**, 1193.
57. A. M. Sescu, L. Favier, D. Lutic, N. Soto-Donoso, G. Ciobanu and M. Harja, TiO_2 Doped with Noble Metals as an Efficient Solution for the Photodegradation of Hazardous Organic Water Pollutants at Ambient Conditions, *Water*, 2021, **13**, 19.
58. Y. Cong, J. Zhang, F. Chen and M. Anpo, Synthesis and Characterization of Nitrogen-Doped TiO_2 Nanophotocatalyst with High Visible Light Activity, *J. Phys. Chem. C*, 2007, **111**, 6976.
59. T. Suwannaruang, P. Kidkhunthod, N. Chanlek, S. Soontaranon and K. Wantala, High anatase purity of nitrogen-doped TiO_2 nanorice particles for the photocatalytic treatment activity of pharmaceutical wastewater, *Appl. Surf. Sci.*, 2019, **478**, 1.
60. S. Leong, A. Razmjou, K. Wang, K. Hapgood, X. Zhang and H. Wang, TiO_2 based photocatalytic membranes: A review, *J. Membr. Sci.*, 2014, **472**, 167.
61. A. W. Zularisam, A. F. Ismail, M. R. Salim, M. Sakinah and O. Hiroaki, Fabrication, fouling and foulant analyses of asymmetric polysulfone (PSF) ultrafiltration membrane fouled with natural organic matter (NOM) source waters, *J. Membr. Sci.*, 2007, **299**, 97.
62. Y. Yang, P. Wang and Q. Zheng, Preparation and properties of polysulfone/TiO_2 composite ultrafiltration membranes, *J. Polym. Sci., Part B: Polym. Phys.*, 2006, **44**, 879.
63. E. M. V. Hoek, A. K. Ghosh, X. Huang, M. Liong and J. I. Zink, Physical–chemical properties, separation performance, and fouling resistance of mixed-matrix ultrafiltration membranes, *Desalination*, 2011, **283**, 89.
64. A. Rahimpour, M. Jahanshahi, B. Rajaeian and M. Rahimnejad, TiO_2 entrapped nano-composite PVDF/SPES membranes: Preparation, characterization, antifouling and antibacterial properties, *Desalination*, 2011, **278**, 343.
65. J. Kim and B. V. d. Bruggen, The use of nanoparticles in polymeric and ceramic membrane structures: Review of manufacturing procedures and performance improvement for water treatment, *Environ. Pollut.*, 2010, **158**, 2335.
66. A. T. Kuvarega, N. Khumalo, D. Dlamini and B. B. Mamba, Polysulfone/N,Pd co-doped TiO_2 composite membranes for photocatalytic dye degradation, *Sep. Purif. Technol.*, 2018, **191**, 122.
67. H. Song, J. Shao, Y. He, B. Liu and X. Zhong, Natural organic matter removal and flux decline with PEG–TiO_2-doped PVDF membranes by integration of ultrafiltration with photocatalysis, *J. Membr. Sci.*, 2012, **405**, 48.

68. D. He, H. Susanto and M. Ulbricht, Photo-irradiation for preparation, modification and stimulation of polymeric membranes, *Prog. Polym. Sci.*, 2009, **34**, 62.
69. D. Emadzadeh, W. J. Lau, T. Matsuura, M. Rahbari-Sisakht and A. F. Ismail, A novel thin film composite forward osmosis membrane prepared from PSf–TiO_2 nanocomposite substrate for water desalination, *Chem. Eng. J.*, 2014, **237**, 70.
70. S. H. Mohamad, M. I. Idris and H. Z. Abdullah, Preparation of Polyethersulfone Ultrafiltration Membrane Surface Coated with TiO_2 Nanoparticles and Irradiated under UV Light, *Key Eng. Mater.*, 2014, **594**, 877.
71. N. A. A. Hamid, A. F. Ismail, T. Matsuura, A. W. Zularisam, W. J. Lau and E. Yuliwati, *et al.*, Morphological and separation performance study of polysulfone/titanium dioxide (PSF/TiO_2) ultrafiltration membranes for humic acid removal, *Desalination*, 2011, **273**, 85.
72. M. S. Jyothi, M. Padaki, R. G. Balakrishna and R. K. Pai, Synthesis and design of PSf/TiO_2 composite membranes for reduction of chromium (VI): Stability and reuse of the product and the process, *J. Mater. Res.*, 2014, **29**, 1537.
73. V. Yusà, C. Coscollà, W. Mellouki, A. Pastor and M. de la Guardia, Sampling and analysis of pesticides in ambient air, *J. Chromatogr. A*, 2009, **1216**, 2972.
74. A. Akyol and M. Bayramoğlu, Photocatalytic degradation of Remazol Red using ZnO catalyst, *J. Hazard. Mater.*, 2005, **124**, 241.
75. B. Pare, S. Jonnalagadda, H. Tomar, P. Singh and V. Bhagwat, ZnO assisted photocatalytic degradation of acridine orange in aqueous solution using visible irradiation, *Desalination*, 2008, **232**, 80.
76. U. I. Gaya, A. H. Abdullah, M. Z. Hussein and Z. Zainal, Photocatalytic removal of 2,4,6-trichlorophenol from water exploiting commercial ZnO powder, *Desalination*, 2010, **263**, 176.
77. A. R. Khataee, M. Zarei, M. Fathinia and M. K. Jafari, Photocatalytic degradation of an anthraquinone dye on immobilized TiO_2 nanoparticles in a rectangular reactor: Destruction pathway and response surface approach, *Desalination*, 2011, **268**, 126.
78. A. Fadaei and M. Sadeghi, Photocatalytic degradation of chlorpyrifos in water using titanium dioxide and zinc oxide, *Fresenius Environ. Bull.*, 2014, **22**, 2442.
79. Gh. Ghanizadeh and G. Asgari, Adsorption kinetics and isotherm of methylene blue and its removal from aqueous solution using bone charcoal, *React. Kinet., Mech. Catal.*, 2011, **102**, 127.
80. X. He, K. B. Male, P. N. Nesterenko, D. Brabazon, B. Paull and J. H. T. Luong, Adsorption and Desorption of Methylene Blue on Porous Carbon Monoliths and Nanocrystalline Cellulose, *ACS Appl. Mater. Interfaces*, 2013, **5**, 8796.
81. Q. Du, J. Sun, Y. Li, X. Yang, X. Wang and Z. Wang, *et al.*, Highly enhanced adsorption of congo red onto graphene oxide/chitosan fibers by wet-chemical etching off silica nanoparticles, *Chem. Eng. J.*, 2014, **245**, 99.

82. L. Wang, Aqueous organic dye discoloration induced by contact glow discharge electrolysis, *J. Hazard. Mater.*, 2009, **171**, 577.
83. M. M. Khan, S. F. Adil and A. Al-Mayouf, Metal oxides as photocatalysts, *J. Saudi Chem. Soc.*, 2015, **19**, 462.
84. M. G. Neelavannan, M. Revathi and C. Ahmed Basha, Photocatalytic and electrochemical combined treatment of textile wash water, *J. Hazard. Mater.*, 2007, **149**, 371.
85. E. Jafarnejad and J. Nemati, Methylene Blue Dye Removal from Aqueous Solutions and Photocatalytic Activity Using UV/Nano-TiO_2 : Operating Parameters Study, *Anal. Chem. Lett.*, 2015, **5**, 192.
86. C. G. Joseph and L. Elilarasi, Removal of Methylene Blue Dye from Aqueous Solution Using a Newly Synthesized TiO_2-SiO_2 Photocatalyst in the Presence of Active Chlorine Species, *IOP Conf. Ser.: Mater. Sci. Eng.*, 2017, **206**, 012090.
87. Z. Pei, Z. Kaiqiang, D. Yu, B. Bo, G. Weisheng and S. Yourui, Adsorption of Organic Dyes by TiO_2@Yeast-Carbon Composite Microspheres and Their In Situ Regeneration Evaluation, *J. Nanomater.*, 2015, **2015**, 498304.
88. W. Chen, H. Xiao, H. Xu, T. Ding and Y. Gu, Photodegradation of Methylene Blue by TiO_2-Fe_3O_4-Bentonite Magnetic Nanocomposite, *Int. J. Photoenergy*, 2015, **2015**, 591428.
89. K. Tolić, M. Runje, T. Gazivoda Kraljević and D. Mutavdžić Pavlović, Identification of Crizotinib Major Degradation Products Obtained Under Stress Conditions by RP-UHPLC-HRMS, *Croat. Chem. Acta*, 2021, **94**, 17.
90. A. Chatzimpaloglou, C. Christophoridis, I. Fountoulakis, M. Antonopoulou, D. Vlastos and A. Bais, *et al.*, Photolytic and photocatalytic degradation of antineoplastic drug irinotecan, Kinetic study, identification of transformation products and toxicity evaluation, *Chem. Eng. J.*, 2021, **405**, 126866.
91. A. de Santiago-Martín, R. Meffe, G. Teijón, V. Martínez Hernández, I. López-Heras and C. Alonso Alonso, *et al.*, Pharmaceuticals and trace metals in the surface water used for crop irrigation: Risk to health or natural attenuation?, *Sci. Total Environ.*, 2020, **705**, 135825.
92. O. R. Fonseca-Cervantes, A. Pérez-Larios, V. H. Romero Arellano, B. Sulbaran-Rangel and C. A. Guzmán González, Effects in Band Gap for Photocatalysis in TiO_2 Support by Adding Gold and Ruthenium, *Processes*, 2020, **8**, 1032.
93. S. A. Al-Zahrani, K. Umar, S. A. Tweib, J. A. M. Rashd, S. K. Afridi and S. A. Bhawani, *et al.*, Biomass Mediated Synthesis of ZnO and ZnO/GO for the Decolorization of Methylene Blue under Visible Light Source, *Catalysts*, 2023, **13**, 409.
94. J. K. Kim, S. D. Kim, J. Y. Lee, C. H. Kim, H. S. Lee and S. M. Koo, *et al.*, A Study on TiO_2 Surface Texturing Effect for the Enhancement of Photocatalytic Reaction in a Total Phosphorous Concentration Measurement System, *Micromachines*, 2021, **12**, 1163.
95. K. Umar, M. M. Haque, M. Muneer, T. Harada and M. Matsumura, Mo, Mn and La doped TiO_2: Synthesis, characterization and photocatalytic

activity for the decolourization of three different chromophoric dyes, *J. Alloys Compd.*, 2013, **578**, 431.

96. A. Rachel, M. Subrahmanyam and P. Boule, Comparison of photocatalytic efficiencies of TiO_2 in suspended and immobilised form for the photocatalytic degradation of nitrobenzenesulfonic acids, *Appl. Catal., B*, 2002, **37**, 301.
97. J. Trawiński and R. Skibiński, Rapid degradation of clozapine by heterogeneous photocatalysis, Comparison with direct photolysis, kinetics, identification of transformation products and scavenger study, *Sci. Total Environ.*, 2019, **665**, 557.
98. H. Mihçiokur, Environmental risk assessment of commonly used anti-cancer drugs, *Cumhuriyet Sci. J.*, 2021, **42**, 310.
99. K. Tolić Čop, D. Mutavdžić Pavlović and T. Gazivoda Kraljević, Photocatalytic Activity of TiO_2 for the Degradation of Anticancer Drugs, *Nanomaterials*, 2022, **12**, 3532.

CHAPTER 8

Application of Nanoparticles in the Mitigation of Harmful Algal Blooms

A. IQBAL[a*], D. H. Y. YANTO[b], N. MOHAMMAD-NOOR[c], H. THOHA[d], M. W. ISMAIL[e], N. H. H. ABU BAKAR[a], M. R. ROZIAWATI[f], S. ABU-ROMMAN[g] AND M. A. SWEISS[g]

[a]School of Chemical Sciences, Universiti Sains Malaysia, 11800 USM, Penang, Malaysia; [b]Research Center for Applied Microbiology, National Research and Innovation Agency (BRIN), Cibinong, Bogor 16911, Indonesia; [c]Institute of Oceanography and Maritime Study, Kulliyyah of Science, International Islamic University Malaysia, Jalan Sultan Ahmad Shah, 25200, Kuantan, Pahang; [d]Research Center for Oceanography, National Research and Innovation Agency (RCO-BRIN), Jl. Pasir Putih I No. 1, Ancol Timur, Jakarta Utara, 14430, Indonesia; [e]Department of Chemistry, Kulliyyah of Science, International Islamic University Malaysia, 25200 Kuantan, Pahang, Malaysia; [f]Fisheries Research Institute, Department of Fisheries Malaysia, 11960 Batu Maung, Penang, Malaysia; [g]Department of Biotechnology, Faculty of Agricultural Technology, Al-Balqa Applied University, Al-Salt, 19117, Jordan
E-mail: anwariqbal@usm.my

8.1 Introduction

Using nanoparticles to combat harmful algal blooms (HABs) is a cutting-edge application that combines the intricacies of nanotechnology with the values of environmental stewardship. An extensive review of the challenges posed

Chemistry in the Environment Series No. 14
Advanced Materials for Emerging Water Pollutant Removal
Edited by Pei Sean Goh, Devagi Kanakaraju, Anwar Iqbal and Ahmad Fauzi Ismail

Published by the Royal Society of Chemistry, www.rsc.org

by HABs, including regional variations in their occurrence, is given in this chapter. In order to treat HABs using nanoparticle technology, it explores the basic ideas of physical and chemical techniques, focusing on adsorption and photocatalysis. In-depth conversations clarify how these strategies use nanoparticles to lessen the effects of HABs successfully.

8.2 Harmful Algal Bloom Definition and Challenges

HABs refer to the rapid growth of microalgal populations, such as cyanobacteria, diatoms, and dinoflagellates, in aquatic ecosystems.[1] The bloom can take place in various types of geographical locations of water bodies such as freshwater, estuarine, and seawater. A single occasion of a bloom can last from days to years, and patches of algae covering a vast area can be seen.[2] It is indisputable that the frequency, duration, and severity of HABS are high due to global warming and excess nutrients from anthropogenic activities such as agriculture, industrialization, urbanization, and population growth.[3]

During algal blooming, the high biomass production limits sunlight penetration for the respiration of aquatic organisms (limiting sunlight for photosynthesis of underwater plants) and can kill fish by causing mechanical damage to the gills. The biomass also hinders water-related activities such as kayaking and swimming. At the end of the bloom season, the decomposition of the biomass uses the dissolved oxygen excessively, leading to hypoxic or anoxic conditions. These conditions are not suitable for the survival of aquatic organisms. In addition, the decomposition also creates a foul smell. Some HAB species produce potent toxins that are lethal to humans and animals when they are exposed to them through the consumption of toxin-contaminated seafood, drinking water, and inhaling the aerosol of the toxin. The economic loss due to HABs can lead to a substantial economic loss, especially to aquaculture operators and tourism industries.

8.3 Occurrences of HABs in Various Regions

HAB is a natural phenomenon that has been reported in many countries around the world. Data on the occurrences of harmful microalgal species have been recorded in the Ocean Biodiversity Information System (OBIS) and in the Harmful Algae Database (HAEDAT). Analysis from these systems showed that a total of 35% of the seafood toxin syndrome is due to paralytic shellfish poisoning (PSP) followed by diarrhetic shellfish poisoning (DSP) (30%), which stems from the toxins of HABs. The paralytic shellfish toxin is mostly reported in North America, the Caribbean, South America, Southeast Asia, and North Asia. In comparison, DSP is highly recorded in Europe and the Mediterranean.

The occurrences of HABs are attributed to two main factors, *i.e.* anthropogenic activities and climate change. In Asian countries such as the Philippines, Malaysia, Hong Kong, Japan, Indonesia, and Korea, numerous new HAB events have been recorded, causing significant impacts to the economy.

Among the common effects of HABs are fish kills, particularly for captive fish in the aquaculture industry. In Malaysia and the Philippines, the type and species of HABs have diversified. Species that have been recorded to cause frequent HAB problems and kill captive fish are *Prorocentrum cordatum*, *Margalefidinium polykrikoides*, *Chattonella* spp., *Karlodinium australe* and *Takayama* sp.[4] As well as causing a significant impact on the aquaculture industry, HABs are also known to be the causative organisms for human illnesses. It has been reported that a total of 105 dinoflagellates, 37 marine bluegreen, 31 diatoms, 8 haptophytes, 6 raphidophytes, 3 dictyochophytes, and 2 pelagophytes can produce toxins.[5]

8.4 Strategic Management of HABs

The strategic management of HABs can be divided into (i) prevention measures to avoid the occurrence of HABs, (ii) control measures to reduce or contain the bloom population, and (iii) mitigation-measures to minimize the ongoing HAB impact on human health, living resources, and coastal economies during the HAB event.[6]

This chapter emphasizes strategic management in terms of mitigation measures. Mitigation refers to any measures employed to reduce the negative impacts of existing or ongoing blooms. Mitigation of HABs can be divided into physical, chemical, and biological approaches. The physical approach refers to any method to physically remove algae cells and their toxins from the water column. Examples of physical approaches are skimming,[7] isolation, ultrasonic destruction,[8] electrolysation of seawater, and sediment-based methods such as nanoparticle flocculation, sediment resuspension, and burial.[9] The chemical approach involves using synthetic or organic chemical compounds from natural resources to interfere with the physiological processes of the algae cells and cause their death. Under the biological process, organic organisms such as seaweed, bacteria, viruses, certain types of fish, *etc.*, are used to eliminate the HABs. Each of these approaches has advantages and disadvantages. Hence, environmental risk assessment on the marine ecosystem must be carried out before implementation in the field.

8.5 Nanoparticles

Nanoparticles are materials in the range 1–100 nm, with a substantial surface-area-to-volume ratio compared to their larger counterparts.[10] This characteristic renders them suitable for various applications, including biomedical, agricultural, food packaging, and energy-related purposes.[11–13] Nanoparticles exhibit distinctive physical and chemical properties, including mechanical, quantum, optical, and electrical characteristics, with silver, zinc oxide, iron, and graphene among the most commonly utilized nanomaterials.[14,15]

The production of nanoparticles can be achieved through either top-down or bottom-up methods. Currently, conventional physical and chemical approaches for synthesizing nanoparticles, such as aerosols, ultraviolet radiation, and thermal decomposition, are gradually being replaced by green synthesis techniques.[16] These green methods may involve the utilization of microorganisms (fungi and bacteria) or plant components (leaves, roots, flowers), resulting in more efficient and environmentally friendly processes.[17]

Nowadays, significant research into the impact of nanoparticles on algae has garnered considerable interest, primarily focusing on their potential to control or restrain algal populations using various mechanisms, including oxidation, photocatalysis, adsorption, and flocculation.[18,19] The exposure of algae to nanoparticles is known to trigger the generation of reactive oxygen species, deplete essential nutrients, and reduce the photosynthetic yield.[20] However, there are still several challenges and risks linked to nanoparticle utilization for mitigating algae, which include their toxicity, potential for bioaccumulation, and potential ecological repercussions.

8.6 Basic Concept of Mitigating HABs Using Nanoparticles

The mitigation of HABs using nanoparticles can be achieved through adsorption or photocatalysis. Applying nanoparticles in HABs mitigation in the field is easy and does not require any maintenance. However, a huge number of nanoparticles are required, which makes this approach relatively expensive. Another drawback of the application of nanoparticles in HABs mitigation is sedimentation. The sedimentation of HABs and its toxins, together with the nanoparticles, could be potentially detrimental to the benthic community.

8.6.1 Physical Method by Adsorption

Adsorption is based on the electrostatic attraction between nanoparticles and the algae. Hence, in addition to the structure and type, the ability of the nanoparticles to be used in mitigating algal cells depends on the electrostatic attraction between the nanoparticles and the algae cells. Since algae are negatively charged in nature,[21] the nanoparticles must be positively charged to attract the algae and form nanoparticle/algae composites.[4] Natural clays, for example, are negatively charged. The algae removal efficiency will be low due to the dominating electrostatic repulsion. Hence, the natural clays are modified with positively charged organic or inorganic linkers. After modification, the overall charge of the modified clays will be less negatively charged, attracting algae for adsorption and removal. The process is simplified in Figure 8.1.

The algae-modified clay composite will then flocculate to form larger, heavier particles or flocs and eventually sediment at the bottom of the water bodies. The interaction mechanism is simplified in Figure 8.2.

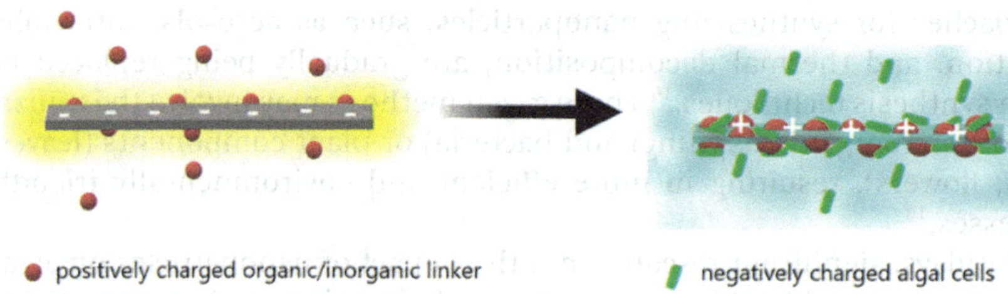

Figure 8.1 Surface modification of negatively charged natural clays with positively charged organic/inorganic linkers increases the positive charge of the composite.

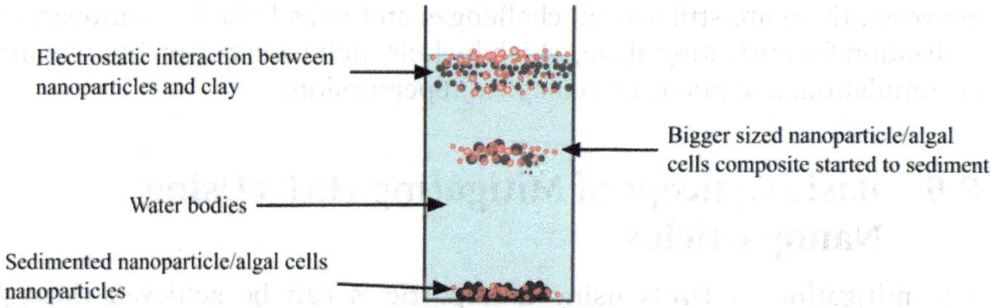

Figure 8.2 Negatively charged algae cells will be attracted to the positively charged nanoparticles, forming larger and heavier particles or flocs. The heavier flocs will eventually sediment.

8.6.1.1 *Clays and Silica-based Adsorbents*

Clays as HABs mitigation agents were first introduced in the late 1970s in the Kagoshima coastal waters, Japan.[22] As mentioned before, clays in their natural form are negatively charged due to surface hydration and lattice defects in seawater.[22] The electrostatic repulsion between the negatively charged clays and algal cells will prevent them from flocculating the electrostatic repulsion on the ineffectiveness of natural clays or silica in mitigating algae. Modification of the clay's surface is an effective method to increase the clay's positive charge and improve flocculation.[23]

Polyaluminium chloride (PAC) is among the frequently used inorganic modifiers. PAC has been declared safe for human health and application in anti-perspirants up to 25%.[24] Zang *et al.* modified kaolinite with PAC using a one-pot method, in which the clay was stirred together with the PAC in hyper-pure water.[25] The modified kaolinite was used to mitigate *Aureococcus anophagefferens*, *Prorocentrum donghaiense*, *Karenia mikimotoi*, and *Nannochloropsis* sp. The group observed that smaller-sized modified clays have strong van der Waals attraction, which causes agglomeration and, subsequently, self-flocculation. In addition, the compression of the double electric layer by salt ions caused the weak electrostatic repulsion between clay particles in seawater. Self-flocculation has a negative impact on the removal of

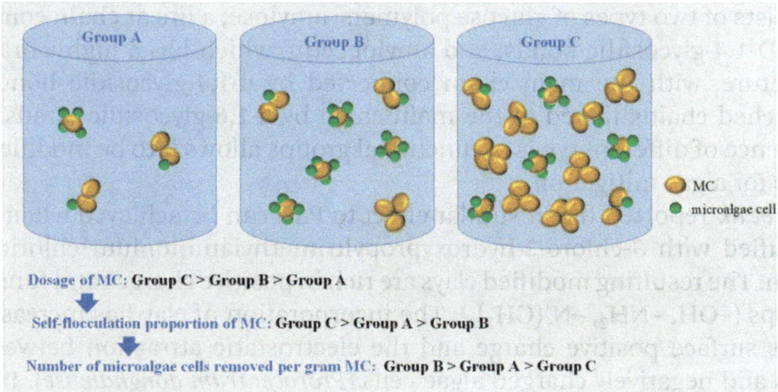

Figure 8.3 Schematic diagram illustrating the influence of modified clay dosage on microalgae removal. Reproduced from ref. 25 with permission from Elsevier, Copyright 2024.

algae cells. The collision rate between the bigger particles and the algae cells was lower compared to smaller particles.[25–28] The influence of modified clay dosage on microalgae removal prepared by the group is summarized in the schematic diagram shown in Figure 8.3.

The modified clays have been identified to have several other advantages besides the ability to form flocs with the algae cells.

Ren *et al.* investigated the PAC-modified kaolin's ability to mitigate Phaeocystis globose. The modification was carried out using a similar method to that described by Zang *et al.*[25] A higher concentration of the PAC-modified kaolin was noticed to induce severe damage to the cells. The damage interrupted the normal physiological processes and inhibited cell proliferation and colony formation. At lower concentrations, the cells suffered minimum damage. The algae cells enhanced polysaccharide synthesis to promote colony formation and alleviate environmental stress.[29] Liu *et al.* reported that clay modified with PAC could cause partial inactivation of reaction centres (RCs) and block the electron transport chain in *Karenia mikimotoi*.[30] PAC-modified kaolin has also been reported to be able to affect nutrient cycling in addition to mitigating HAB. The modified kaolin is able to remove phosphorous from surface waters and adsorb organic phosphorus on its matrix by forming a stable complex with the aluminium in the clay matrix.[31]

PAC contains aluminium, which can harm the aquatic ecosystem in the long run.[32] Organic linkers such as chitosan is an alternative environmentally friendly option to replace PAC. In addition, chitosan is biodegradable and biocompatible. Chitosan is a natural polysaccharide that is comprised of repeating *N*-acetyl-D-glucosamine and D-glucosamine units.[33] Regardless, chitosan alone is not suitable for use in seawater. The high ionic strength and alkalinity prevent the unfolding of the polymer chain, which weakens the chitosan's netting and bridging properties.[34,35] Starch has similar properties to chitosan that can be developed to replace PAC. Starch is a polysaccharide that consists of two types of glucose polymers: amylose, a linear chain composed of α-D-1,4-glycosidic bonds, and amylopectin, which has a highly branched structure, with the main chain connected by α-1,4-glycosidic bonds and branched chains linked to the main chain by α-1,6-glycosidic bonds.[36] The presence of different types of functional groups allows it to be modified with clays for algae mitigation.

Li *et al.* reported that a similar effect to PAC can be achieved when clay is modified with 3-chloro-2-hydroxypropyltrimethylammonium chloride chitosan. The resulting modified clays are rich in positive charges and functional groups (–OH, $–NH_2$, $–N^+(CH_3)_3$). The incorporation of clay has increased the clay's surface positive charge and the electrostatic attraction between the clays and negatively charged algae cells (*Prorocentrum donghaiense*). In addition, algal cells were also observed to be adsorbed on the molecular chain of modified chitosan with stronger chemical bonds. At a dosage of 20 mg L^{-1}, the removal efficiency of *Prorocentrum donghaiense* was approximately 80%. Similar to PAC-modified clays, the HABs mitigation using chitosan-modified clays takes place according to the flocculation–sedimentation approach.[37]

Xia and colleagues reported the modification of attapulgite (ATP) with starch-3-chloro-2-hydroxypropyltrimethylammonium chloride to mitigate *Microcystis aeruginosa*.[38] Since the crystallinity of the ATP did not change, it is deduced that the starch was grafted on the surface. The algae cell removal mechanism is through complete charge-neutralization, bridging, and sweeping effects, as shown in Figure 8.4. The presence of Mg^{2+}, Ca^{2+}, and Fe^{3+} in the algal cultural medium was hydrolysed to form metal hydroxides, contributing to the effective coagulation.[39] Typically, ammonium groups may cause

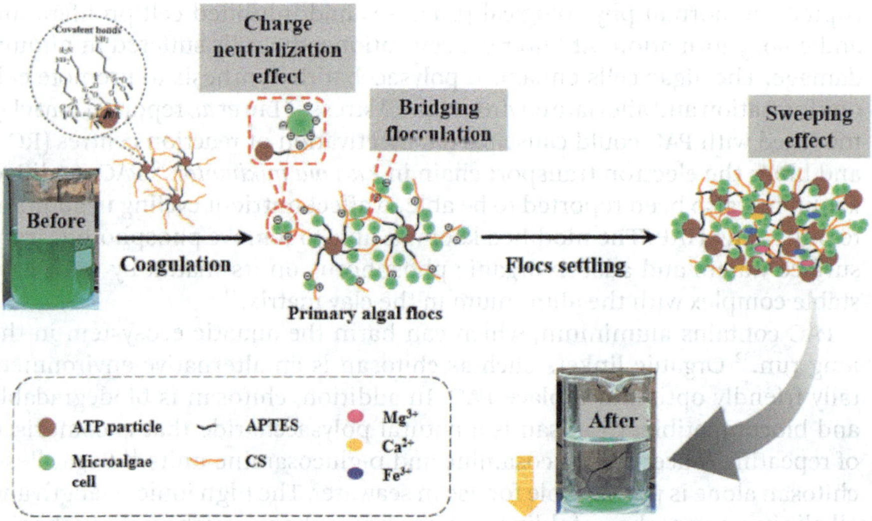

Figure 8.4 The coagulation mechanisms of algae using the modified ATP. Reproduced from ref. 38 with permission from Elsevier, Copyright 2024.

algae cellular rupture with consequent toxin release. However, the group reported that significant cell rupture did not take place.[38]

Soil and sand have also been used and modified as an alternative to clays. However, sands have lesser flocculating ability compared to natural clays. Pan *et al.* reported that the flocculating ability of sand could be improved by adding a small amount of PAC and chitosan to mitigate *Amphidinium carterae* Hulburt and *Chlorella* sp. The PAC assisted in keeping the netting and bridging function of chitosan in seawater and in forming small flocs, whereas the chitosan gathered the small flocs into large and denser flocs.[22]

Jin *et al.* modified soil using amphoteric starch (AS) and PAC to mitigate *Microcystis aeruginosa* and marine *Chlorella* sp. The AS-PAC modified soils exhibit resistance to salinity change due to the anti-polyelectrolyte effect of AS. As a result, 99.9% removal efficiency was achieved for *M. aeruginosa* in fresh and saline waters and *Chlorella* sp. in marine water, respectively.[40] The positive and negative groups in the molecular chain of the AS attracted both positive and negative flocs.[41]

Plant tannins are the fourth most abundant forest byproducts after cellulose, lignin, and hemicellulose.[42] Despite its low isoelectric point ($pH_{pzc} \sim 2.0$), the polyphenolic hydroxyl groups in tannins make it suitable for developing flocculants.[43] Chen and Pan reported that 55% of *Amphidinium carterae* cells were removed by xanthan alone. The removal increased to 83–89% within 30 min using 300 mg L^{-1} clays, soils, or sands modified with 20 mg L^{-1} xanthan and 100 mg L^{-1} calcium hydroxide. The xanthan could trap and wrap algal cells through bridging and netting interactions due to its superior salt compatibility in seawater.[44]

8.6.2 Photocatalysis

Photocatalysts are semiconductors that change the course of reactions in the presence of light irradiation. The process is known as photocatalysis. During the photocatalysis process, electrons (e^-) will be excited from the valence band (VB) to the conductive band (CB), leaving behind positively charged holes (h^+) in the VB. The photogenerated e^-/h^+ pairs migrate to the photocatalyst's surface and react with the surface adsorbed oxygen and water (or OH^-) to form reactive oxygen species (ROS) such as superoxide radical ($O_2^{\bullet-}$), singlet oxygen ($^1O_2$), hydroxyl radical ($O^\bullet$), hydroperoxyl radical ($HOO^\bullet$) and hydroperoxide (H_2O_2).[45] The photocatalytic process is simplified in Figure 8.5.

8.6.2.1 Photocatalysis and HABs Mitigation

Photocatalysis is commonly used to mineralize organic compounds such as dyes, pesticides, and antibiotics. The potential of photocatalysis in mitigating HABs has attracted the interest of many researchers in the last few years.

Unlike clays, soils, and sand, photocatalysts have unique ways of deactivating or killing the HAB. Previous studies have shown that photocatalysts can cause oxidative stress due to the generation of ROS mainly[46,47] and through

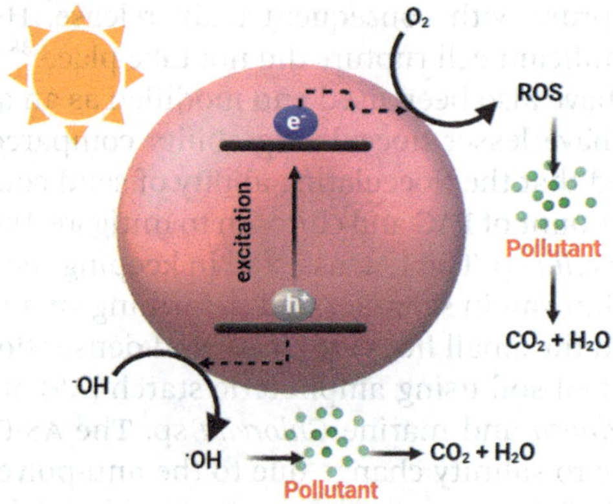

Figure 8.5 General photocatalysis mechanism.

physical constraints (shading effect[48]), aggregation (homoagglomeration and heteroagglomeration), sedimentation effect, and direct membrane and cellular structure damages.[49,50]

8.6.2.2 Photocatalysts for the Mitigation of HABs

Various types of photocatalysts have been developed to mitigate HABs. However, only a few examples will be discussed. Photocatalysis faces similar drawbacks to adsorption. A large number are required for mitigation and the photocatalysts are prone to sedimentation. In addition, the disadvantage of ROS is that they are not species-selective and can cause distress to other organic organisms.

Lee *et al.*[51] developed a buoyant structure prototype containing TiO_2-embedded on expanded polystyrene (TiO_2EPS) balls to mitigate HABs in the field. An *in situ* real-time monitoring device (Algaecheck, Modern Water, UK) was installed inside the floating structure to measure the real-time changes in concentrations of both Chlorophyll a (Chl-a) and phycocyanin in water. The structure is shown in Figure 8.6.

The prototype was placed at the Hallyu stream located in Goyang-si, Gyeonggi-do, Republic of Korea. The Hallyu stream contains pollutants from urban and rural areas and anthropogenic sources. This method used continuous solar irradiation and oxygenation in the air–water interface to adsorb the HAB cells and degrade the cells into the intermediates by photocatalytic activity through the embedded TiO_2 nanoparticles. The balls also created shaded areas preventing sunlight from illuminating/reaching the water column to prevent the growth and reproduction of HABs. The effectiveness of the balls in controlling the growth of the HAB depends on the solar irradiation levels, water depth, climate, nutrient influx, *etc.* and requires yearly replacement to maintain its effectiveness.

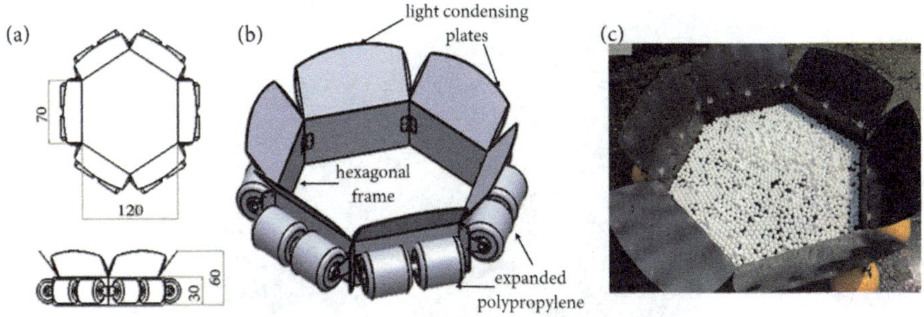

Figure 8.6 TiO_2EPS balls mesocosm with hexagonal stainless-steel structure and light condensing plates floating on the hypereutrophic water body for field application. (a) Plan and cross-section views (unit: cm), (b) 3D rendering, and (c) pictorial view. Reproduced from ref. 51 with permission from Elsevier, Copyright 2024.

Nitrogen and phosphorous-doped TiO_2/C composites were developed by Wang *et al.* to remove *Microcystis aeruginosa via* an adsorption–photocatalytic approach. The composite's rough surface and porous structure facilitated the algae cells' adsorption. The total removal of algae cells was 92.6% under visible light irradiation for 6 h. The generated ROS were reported to be able to interrupt the metabolic activities of the algae, leading to its death. The algae cells were observed to be adsorbed on the catalyst surfaces. The edges of the cells were malformed, and the colour of the cells became lighter. The changes in the colour indicate either the loss or breakdown of the chlorophyll. Furthermore, the composite can effectively degrade the cytotoxins from water.[52]

Kennedy *et al.* reported the ability of 3D printed TiO_2 composites to focus on removing microcystin produced by *M. aeruginosa* UTEX 2385. The composite material reduced the microcystin cyanotoxin ($t_{1/2}$ = 3 h) faster than photolysis ($t_{1/2}$ = 36 h). However, the process is slower in the presence of the cyanobacteria cells ($t_{1/2}$ = 11 h) due to cell activity and light attenuation. The 3D-printed TiO_2 composite is shown in Figure 8.7. The mechanism of photocatalytic degradation and mineralization of microcystin cyanotoxin is attributed to $OH^{\bullet}$ attack and diene bond and hydroxylation.[53]

Fan *et al.*, reported the synthesis of plasmonic Ag/AgCl@$LaFeO_3$ photocatalyst *via* an *in situ* method for removing *Microcystis aeruginosa* under visible light irradiation.[54] The Ag/AgCl@$LaFeO_3$ has a better light absorption capacity and lower photogenerated electron–hole recombination rate compared to pure $LaFeO_3$. The removal rate of Chl-a was almost 100% within 150 min. The kinetic constants were 15.36 and 9.61 times faster than $LaFeO_3$ and Ag/AgCl, respectively. The photocatalyst changed the permeability and morphology of the algae cell membrane and destroyed the photosynthesis and antioxidant systems of *M. aeruginosa*. The ROS responsible for the degradation was identified as $O_2^{\bullet-}$ and $OH^{\bullet}$. The valence band edge potential (E_{VB}) of $LaFO_3$ and AgCl were 2.19 eV and 3.01 eV, respectively. Meanwhile, the conductive band edge potential (E_{CB}) was 0.17 eV and −0.09 eV for $LaFO_3$ and AgCl, respectively.

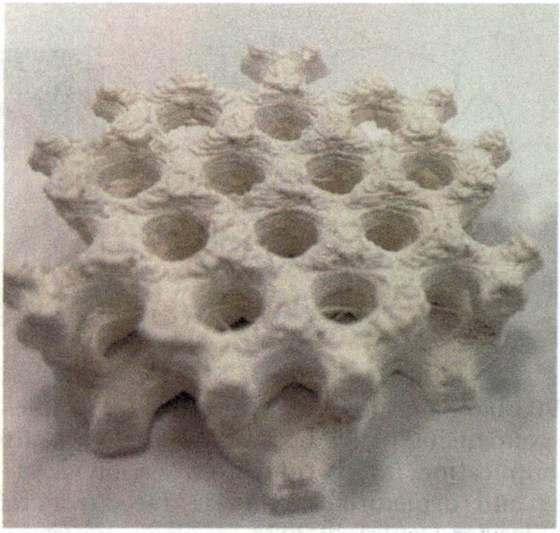

Figure 8.7 3D printed TiO_2 composite. Reproduced from ref. 53 with permission from Elsevier, Copyright 2024.

The narrow band gap of $LaFO_3$ (2.02 eV) excites the electrons by visible light to the CB. However, this excitation cannot occur for AgCl due to its wide band gap (3.10 eV). The metallic Ag^0 is able to absorb visible light and generate electron–hole pairs due to the surface plasmon resonance (SPR) effect. The photogenerated electrons on the CB band of $LaFeO_3$ were transferred to the lowest empty orbital of Ag^0, reducing the recombination rate of photogenerated electron–hole pairs. In addition, the electrons transferred to Ag^0 can increase its Fermi level.[55] Since the CB potential of AgCl (−0.09 eV) is more negative than the redox potential of O2/ $O_2^{\bullet-}$ (−0.046 eV), the dissolved O_2 in the water can trap the electrons on the AgCl surface to form $O_2^{\bullet-}$.[56] The holes in the VB of $LaFeO_3$ can oxidize OH− to OH$^{\bullet}$ due to the more positive potential of the VB in $LaFeO_3$ (2.19 eV) than that of OH−/OH$^{\bullet}$ (1.99 eV).[57] These radicals will be continuously produced for the photodegradation.[58]

The application of magnetic nanoparticles in HAB mitigation is an interesting approach to explore. The main advantage of magnetized nanoparticles, compared to the other types of photocatalysts, is the ease of separation using a magnet. However, this technology's feasibility and safety with other aquatic organism in the field is questionable. Magnetic nanocomposites have been found to have positive outcomes in HAB mitigation, especially regarding separation. A novel co-graft tannin and polyethyleneimine co-coating magnetic composite (TP@Fe_3O_4) was synthesised by Yang *et al.* for the mitigation of *M. aeruginosa* and algal organic matter (AOM). The approach was based on magnetic flocculation and sulfate radical ($SO_4^{\bullet-}$) advanced oxidation processes (S-AOPs).[59] The composite removed >99% algae cells at a dosage of 100 mg L^{-1} at pH = 8.0 within 5 min. The cell stability was lowered, and the cell flocculation was promoted through peroxymonosulfate (PMS) treatment. The ultraviolet (UV) irradiation activated PMS for the total degradation of AOM and microcystin. The $SO_4^{\bullet -}$ was identified as the main radical species responsible for the removal.[60]

A novel perovskite-type catalyst La_2CuO_4/PMS was reported to be able to remove *M. aeruginosa* cells by Gao *et al.* The removal of *M. aeruginosa* cells and Chl-a was 97% and 100%, respectively, within 20 min and dissolved organic carbon and metals were not released in water. The $OH^{\bullet}$, $SO_4^{\bullet -}$, $^1O_2$ and $O_2^{\bullet -}$ were identified as responsible for the algae inactivation and subsequent algal organic matter removal. The La is proposed to destabilize the algae suspension by increasing its zeta potential, whereas the Cu facilitated the electron transfer to improve the production of $OH^{\bullet}$ and $SO_4^{\bullet -}$.[61]

Metal–organic frameworks (MOFs) contain metal subunits and multidentate organic linkers bonded with coordination bonds. Due to their abundant coordinated unsaturated metal sites, high specific surface area, and visible light response, MOFs have a high potential to be used as photocatalysts.[62,63] Wang *et al.* reported that copper-based MOF (Cu-MOF) exhibit good performance in mitigating *M. aeruginosa via* photocatalysis. The OD_{680} value of algal suspensions decreased to 372% after being exposed to 1 mg L^{-1} Cu-MOF for 24 h.[64] Similar to other photocatalysts, the mechanism of algal inactivation was because Cu-MOF generated free radicals under light irradiation, causing oxidative stress on algae.

8.7 Conclusions

In conclusion, the mitigation of HABs can be addressed using a variety of nanoparticles, either *via* adsorption or photocatalysis. Although, in general, the photocatalytic approach seems more effective in addressing the issue of HABs by the generation of free radicals that can deactivate or kill HABs as well as offering several advantages such as magnetic properties for facile recollection and the possible applicability under visible light, it is crucial to realise that this range of nanoparticles can also affect its surroundings in terms of killing organisms which are important to the ecosystem. Hence, until a more targeted photocatalyst can be designed, the physical approach may be considered as a more practical alternative to photocatalysis. The adsorption of HABs would have a lesser impact on the ecosystem as a whole. It is discerned that the physical approach utilizing nanoparticles can be considered the best, considering its efficiency and practicality.

Acknowledgements

The authors thank Universiti Sains Malaysia for supporting this research through Geran Penyelidikan Jangka Pendek-Jordan (R501-LR-RND002-0000000341-0000) and the Fisheries Research Institute, Department of Fisheries Malaysia, Batu Maung, Penang for the facilities provided. Anwar Iqbal would like to thank the National Research and Innovation Agency (BRIN), Government of the Republic of Indonesia, for the appointment as a visiting researcher during the sabbatical program.

References

1. D. M. Anderson, E. Fensin, C. J. Gobler, A. E. Hoeglund, K. A. Hubbard, D. M. Kulis, J. H. Landsberg, K. A. Lefebvre, P. Provoost, M. L. Richlen, J. L. Smith, A. R. Solow and V. L. Trainer, *Harmful Algae*, 2021, **102**, 101975.
2. J. N. Fink, *Am. Fam. Physician*, 1978, **18**, 124–127.
3. University of Maryland Center for Environmental Science, P. Glibert and M. Burford, *Oceanography*, 2017, **30**, 58–69.
4. A. T. Yñiguez, P. T. Lim, C. P. Leaw, S. J. Jipanin, M. Iwataki, G. Benico and R. V. Azanza, *Harmful Algae*, 2021, **102**, 101776.
5. G. M. Hallegraeff, D. M. Anderson, C. Belin, M.-Y. D. Bottein, E. Bresnan, M. Chinain, H. Enevoldsen, M. Iwataki, B. Karlson, C. H. McKenzie, I. Sunesen, G. C. Pitcher, P. Provoost, A. Richardson, L. Schweibold, P. A. Tester, V. L. Trainer, A. T. Yñiguez and A. Zingone, *Commun. Earth Environ.*, 2021, **2**, 117.
6. D. M. Anderson, *Ocean Coastal Manage.*, 2009, **52**, 342–347.
7. S. Jiang, Y. Li and B. P. Ladewig, *Sci. Total Environ.*, 2017, **595**, 567–583.
8. J. Park, J. Church, Y. Son, K.-T. Kim and W. H. Lee, *Ultrason. Sonochem.*, 2017, **38**, 326–334.
9. A. H. Alshahri, L. Fortunato, N. Ghaffour and T. Leiknes, *Chemosphere*, 2021, **274**, 129676.
10. N. Joudeh and D. Linke, *J. Nanobiotechnol.*, 2022, **20**, 262.
11. S. Kushwaha, A. K. Pandey and N. Kushwala, *Lett. Appl. NanoBioSci.*, 2021, **11**, 4050–4060.
12. A. Valsalan and P. Sivaranjana, *Anal. Methods Environ. Chem. J.*, 2022, **5**, 20–42.
13. M. C. Bruno, M. C. Cristiano, C. Celia, N. d'Avanzo, A. Mancuso, D. Paolino, J. Wolfram and M. Fresta, *ACS Nano*, 2022, **16**, 19665–19690.
14. M. M. Ghazimoradi, F. V. Azad, F. Jalali and M. Rafieian-Kopaei, *Curr. Pharm. Des.*, 2022, **28**, 3572–3581.
15. K. M. F. Hasan, L. Xiaoyi, Z. Shaoqin, P. G. Horváth, M. Bak, L. Bejó, G. Sipos and T. Alpár, *Heliyon*, 2022, **8**, e12322.
16. I. Hussain, N. B. Singh, A. Singh, H. Singh and S. C. Singh, *Biotechnol. Lett.*, 2016, **38**, 545–560.
17. S. Jadoun, R. Arif, N. K. Jangid and R. K. Meena, *Environ. Chem. Lett.*, 2021, **19**, 355–374.
18. Q. Yue, X. He, N. Yan, S. Tian, C. Liu, W.-X. Wang, L. Luo and B. Z. Tang, *Chem. Eng. J.*, 2021, **417**, 127890.
19. B. Ren, K. A. Weitzel, X. Duan, M. N. Nadagouda and D. D. Dionysiou, *Sep. Purif. Technol.*, 2022, **293**, 121106.
20. F. Wang, W. Guan, L. Xu, Z. Ding, H. Ma, A. Ma and N. Terry, *Appl. Sci.*, 2019, **9**, 1534.
21. N. Li, P. Wang, S. Wang, C. Wang, H. Zhou, S. Kapur, J. Zhang and Y. Song, *J. Environ. Chem. Eng.*, 2022, **10**, 107516.
22. G. Pan, J. Chen and D. M. Anderson, *Harmful Algae*, 2011, **10**, 381–387.
23. X. Cao and Z. Yu, *Yingyong Shengtai Xuebao*, 2003, **14**, 1169–1172.

24. A. S. A. Seger and G. Hallegraeff, *Proceedings of the 17th International Conference on Harmful Algae 2016*, 2016, pp. 154–157.
25. X. Zang, Z. Yu, W. Jiang, X. Song and X. Cao, *Sep. Purif. Technol.*, 2023, **305**, 122422.
26. S. Yang, Y. Wu, C. Qu, J. B. Fein, Y. He, Q. Huang and P. Cai, *Colloids Surf., B*, 2021, **198**, 111497.
27. J. B. Morrow, R. Stratton, H.-H. Yang, B. F. Smets and D. Grasso, *Environ. Sci. Technol.*, 2005, **39**, 6395–6404.
28. R.-S. Juang and R.-C. Shiau, *J. Membr. Sci.*, 2000, **165**, 159–167.
29. X. Ren, Z. Yu, X. Song, J. Zhu, W. Wang and X. Cao, *Sci. Total Environ.*, 2022, **838**, 155985.
30. S. Liu, Z. Yu, X. Song and X. Cao, *Mar. Pollut. Bull.*, 2018, **133**, 491–499.
31. G. Lu, X. Song, Z. Yu and X. Cao, *J. Appl. Phycol.*, 2017, **29**, 917–928.
32. E. Gauthier, I. Fortier, F. Courchesne, P. Pepin, J. Mortimer and D. Gauvreau, *Environ. Res.*, 2000, **84**, 234–246.
33. C. Lim, D. S. Hwang and D. W. Lee, *Carbohydr. Polym.*, 2021, **259**, 117782.
34. G. Qun and W. Ajun, *Carbohydr. Polym.*, 2006, **64**, 29–36.
35. H. Zou, G. Pan and H. Chen, *Huanjing Kexue*, 2005, **26**, 148–151.
36. M. S. Dorneles, E. S. De Azevedo and C. P. Z. Noreña, *Int. J. Biol. Macromol.*, 2023, **253**, 126981.
37. H. Li, Z. Yu, X. Cao and X. Song, *Environ. Technol. Innovation*, 2023, **29**, 103028.
38. W. Xia, Y. Wang, S. Shen and H. Yang, *Chem. Eng. J.*, 2023, **470**, 144116.
39. Y. You, L. Yang, X. Sun, H. Chen, H. Wang, N. Wang and S. Li, *J. Cleaner Prod.*, 2022, **344**, 131042.
40. X. Jin, L. Bi, T. Lyu, J. Chen, H. Zhang and G. Pan, *Water Res.*, 2019, **165**, 115005.
41. H. Peng, S. Zhong, Q. Lin, X. Yao, Z. Liang, M. Yang, G. Yin, Q. Liu and H. He, *Carbohydr. Polym.*, 2016, **138**, 210–214.
42. M. Vera and B. F. Urbano, *Polym. Chem.*, 2021, **12**, 4272–4290.
43. Z. Yang, J. Hou, M. Wu, L. Miao, J. Wu and Y. Li, *Sci. Total Environ.*, 2022, **806**, 150518.
44. J. Chen and G. Pan, *J. Appl. Phycol.*, 2012, **24**, 1183–1189.
45. H. Dong, G. Zeng, L. Tang, C. Fan, C. Zhang, X. He and Y. He, *Water Res.*, 2015, **79**, 128–146.
46. F. Li, Z. Liang, X. Zheng, W. Zhao, M. Wu and Z. Wang, *Aquat. Toxicol.*, 2015, **158**, 1–13.
47. N. Von Moos and V. I. Slaveykova, *Nanotoxicology*, 2014, **8**, 605–630.
48. N. B. Hartmann, F. Von Der Kammer, T. Hofmann, M. Baalousha, S. Ottofuelling and A. Baun, *Toxicology*, 2010, **269**, 190–197.
49. M. Sendra, M. P. Yeste, J. M. Gatica, I. Moreno-Garrido and J. Blasco, *Sci. Total Environ.*, 2017, **592**, 403–411.
50. Y. Wang, X. Zhu, Y. Lao, X. Lv, Y. Tao, B. Huang, J. Wang, J. Zhou and Z. Cai, *Sci. Total Environ.*, 2016, **565**, 818–826.
51. S. Lee, C. H. Ahn, E. J. Kim, J. R. Park and J. C. Joo, *J. Hazard. Mater.*, 2020, **398**, 123172.

52. X. Wang, J. Song, J. Zhao, Z. Wang and X. Wang, *Chemosphere*, 2019, **228**, 351–359.
53. A. J. Kennedy, A. D. McQueen, M. L. Ballentine, L. R. May, B. M. Fernando, A. Das, K. L. Klaus, C. B. Williams and M. J. Bortner, *Chem. Eng. J.*, 2023, **455**, 140866.
54. G. Fan, Z. Chen, Z. Yan, B. Du, H. Pang, D. Tang, J. Luo and J. Lin, *J. Hazard. Mater.*, 2021, **409**, 125018.
55. J. W. Ha, T. P. A. Ruberu, R. Han, B. Dong, J. Vela and N. Fang, *J. Am. Chem. Soc.*, 2014, **136**, 1398–1408.
56. S.-F. Yang, C.-G. Niu, D.-W. Huang, H. Zhang, C. Liang and G.-M. Zeng, *Environ. Sci.: Nano*, 2017, **4**, 585–595.
57. Y. Wu, H. Wang, W. Tu, Y. Liu, Y. Z. Tan, X. Yuan and J. W. Chew, *J. Hazard. Mater.*, 2018, **347**, 412–422.
58. G. Fan, Z. Chen, Z. Yan, B. Du, H. Pang, D. Tang, J. Luo and J. Lin, *J. Hazard. Mater.*, 2021, **409**, 125018.
59. Z. Yang, J. Hou, Z. Pan, M. Wu, M. Zhang, J. Wu and L. Miao, *J. Hazard. Mater.*, 2022, **439**, 129485.
60. Z. Yang, J. Hou, Z. Pan, M. Wu, M. Zhang, J. Wu and L. Miao, *J. Hazard. Mater.*, 2022, **439**, 129485.
61. P. Gao, Y. He, S. Lu, M. He, Z. Liu, Y. Deng, Z. Liu, T. Xu and H. Zhang, *Appl. Catal., B*, 2023, **328**, 122511.
62. Z. Liang, C. Qu, D. Xia, R. Zou and Q. Xu, *Angew. Chem., Int. Ed.*, 2018, **57**, 9604–9633.
63. R. Freund, O. Zaremba, G. Arnauts, R. Ameloot, G. Skorupskii, M. Dincă, A. Bavykina, J. Gascon, A. Ejsmont, J. Goscianska, M. Kalmutzki, U. Lächelt, E. Ploetz, C. S. Diercks and S. Wuttke, *Angew. Chem., Int. Ed.*, 2021, **60**, 23975–24001.
64. Z. Wang, Y. Xu, C. Wang, L. Yue, T. Liu, Q. Lan, X. Cao and B. Xing, *Sep. Purif. Technol.*, 2023, **313**, 123515.

Subject Index